20th Annual International Astrophysics Conference 2022

Journal of Physics: Conference Series Volume 2544

Santa Fe, New Mexico, USA
31 October - 1 November 2022

Editor:

G P Zank

ISBN: 978-1-7138-7704-2
ISSN: 1742-6588

Printed from e-media with permission by:

Curran Associates, Inc.
57 Morehouse Lane
Red Hook, NY 12571

Some format issues inherent in the e-media version may also appear in this print version.

Institute of Physics
Dirac House, Temple Back
Bristol BS1 6BE UK

Phone: 44 1 17 929 7481
Fax: 44 1 17 920 0979

techtracking@iop.org

Additional copies of this publication are available from:

Curran Associates, Inc.
57 Morehouse Lane
Red Hook, NY 12571 USA
Phone: 845-758-0400
Fax: 845-758-2633
Email: curran@proceedings.com
Web: www.proceedings.com

TABLE OF CONTENTS

Preface

G P Zank
Editor, *Proceedings of the 20th Annual International Astrophysics Conference,*
Center for Space Plasma and Aeronomic Research (CSPAR) and Department of Space Physics, University of Alabama in Huntsville, Huntsville AL 35805, USA

E-mail: garyp.zank@gmail.com

The 20th Annual International Astrophysics Conference (AIAC) was held at the La Posada de Santa Fe Hotel, Santa Fe, New Mexico, USA from October 31 to November 4, 2022. This year's meeting followed in the footsteps of 2020's COVID-19-affected meeting and was the first AIAC since the pandemic. Understandably, the re-opening of the AIAC meetings proceeded cautiously but we are happy to report that the meeting went exceptionally smoothly, safety protocols were followed, and an enjoyable and stimulating meeting was held. Leading the way in ensuring a safe and successful meeting was of course the conference organizer Adele Corona of ICNS and her assistant Kathy Tolbert, and the staff at the La Posada. With the successful conclusion of this meeting, we look forward to a degree of normalcy in organizing meetings and conferences in future.

The theme of the 20th AIAC was "*From the Depths of the Solar Corona to the Darkness of Interstellar Space: A 20 Year Perspective*" with our now traditional format of 25-minute presentations punctuated by selected 40-minute invited talks that explored various themes in greater detail.

The intent of the 20th AIAC was twofold – retrospective, celebrating the solar, space, and astrophysical accomplishments of the past 20 years as expressed in the 19 prior AIAC conferences, and perspective, i.e., examining some of the pressing questions that taxed our community over the past 20 years from the perspective of what we think we now know and understand. Since the 19 previous meetings addressed the broad themes of the interaction of the outer heliosphere and the local interstellar medium, shock waves and nonlinear processes, energetic particles, particle acceleration and transport, the heating and physics of the solar corona and solar wind, turbulence processes, and much more, these were the themes of the 20th iteration of the AIAC. A characteristic theme of previous meetings was the universality of many of the physical processes that informed the conference, and this was a central element of the 20th celebration.

Finally, let me thank again Adele Corona for another outstanding meeting. Her ability to create and organize superbly run and efficient conferences is unmatched. Likewise, we thank her long-time assistant Kathy Tolbert for her excellent help. Finally, thanks as always to Adele for her help in providing the logistical support for this volume of papers.

Journal of Physics: Conference Series **2544** (2023) 011002 doi:10.1088/1742-6596/2544/1/011002

Peer Review Statement

All papers published in this volume have been reviewed through processes administered by the Editors. Reviews were conducted by expert referees to the professional and scientific standards expected of a proceedings journal published by IOP Publishing.

- **Type of peer review:** Undefined
- **Conference submission management system:** Morressier
- **Number of submissions received:** 9
- **Number of submissions sent for review:** 9
- **Number of submissions accepted:** 11
- **Acceptance Rate (Submissions Accepted / Submissions Received × 100):** 100
- **Average number of reviews per paper:** 1
- **Total number of reviewers involved:** 9
- **Contact person for queries:**
  Name: Adele Corona
  Email: icnsmeetings@gmail.com
  Affiliation: International Networking & Conference Solutions (ICNS)

A Spherical Shells Model of Atmospheric Absorption for Instrument Calibration

N Donders[1,2], G Vigil[3], A Kobelski[3], L J Paxton[4], A Winebarger[3], C Kankelborg[5] and G P Zank[1,2]

[1] The University of Alabama in Huntsville, Huntsville AL
[2] The Department of Space Science and The Center for Space Plasma and Aeronomic Research, Huntsville AL
[3] NASA Marshall Space Flight Center, Huntsville AL
[4] Johns Hopkins Applied Physics Laboratory, Laurel MD
[5] Montana State University, Bozeman MT

E-mail: nicolas.donders@uah.edu, nicolasdonders@gmail.com

Abstract.
We present a model for atmospheric absorption of solar ultraviolet (UV) radiation. The initial motivation for this work is to predict this effect and correct it in Sounding Rocket (SR) experiments. In particular, the Full-sun Ultraviolet Rocket Spectrograph (FURST) is anticipated to launch in mid-2023. FURST has the potential to observe UV absorption while imaging solar spectra between 120-181 nm, at a resolution of $\mathcal{R} > 2 \cdot 10^4$ ($\Delta V < \pm 15$ km/s), and at altitudes of between $\approx$ 110-255 km. This model uses estimates for density and temperature, as well as laboratory measurements of the absorption cross-section, to predict the UV absorption of solar radiation at high altitudes. Refraction correction is discussed and partially implemented but is negligible for the results presented. Absorption by molecular Oxygen is the primary driver within the UV spectral range of our interest. The model is built with a wide range of applications in mind. The primary result is a method for inversion of the absorption cross-section from images obtained during an instrument flight, even if atmospheric observations were not initially intended. The potential to obtain measurements of atmospheric properties is an exciting prospect, especially since sounding rockets are the only method currently available for probing this altitude *in-situ*. Simulation of noisy spectral images along the FURST flight profile is performed using data from the High-Resolution Telescope and Spectrograph (HRTS) SR and the FISM2 model for comparison. Analysis of these simulated signals allows us to capture the Signal-to-Noise Ratio (SNR) of FURST and the capability to measure atmospheric absorption properties as a function of altitude. Based on the prevalence of distinct spectral features, our calculations demonstrate that atmospheric absorption may be used to perform wavelength calibration from in-flight data.

1. Introduction

Atmospheric absorption, in particular from Oxygen and Nitrogen molecules, significantly limits our ability to observe the Ultraviolet (UV) spectrum from the ground [1]. Many teams use sounding rockets, weather balloons, or high-altitude observatories for aeronomy or solar physics studies to get above most of the atmosphere. Correcting for this absorption is possible, however, some light will always be lost.

There are many models that predict the properties of Earth's upper atmosphere. The Naval Research Laboratory Mass Spectrometer and Incoherent Scatter Exosphere 2000 (NRLMSISE-00) atmospheric model is typically used to obtain the density of elements and their temperatures as they vary with altitude [2]. The most common absorption coefficient database, The High-Resolution Transmission (HITRAN) molecular spectroscopic database is a common resource for obtaining absorption coefficients, but it does not contain absorption cross-section information for O_2 in the UV spectral range [3]. The Max-Planck Institute for Chemistry in Mainz, Germany (MPI-Mainz) Ultraviolet-Visible (UV-VIS) database [4–6] does contain this data, however, it requires significant post-processing.

In our research, there are very few complete models for UV absorption and even fewer that cover wavelengths below 300 nm [7–11]. To give an example, [12–14] aims to calculate the high-altitude atmospheric emission effects due to "day-glow." However, it does this by looking through Earth's upper atmospheric layers. This absorption model was designed for the specific purpose of determining the optical depth between two orbiting spacecraft and inferring the abundance of certain molecules in that level of the upper atmosphere. Many aeronomy models account for the self-absorption and re-emission of light, and more information can be found in [15].

A pressing issue with all of these available models is that they are closed-source and platform-specific programs. It would be extremely beneficial to the atmospheric community if a general-purpose program could be developed that pulled together these various resources. Thus, we have developed an open-source platform-independent model with the primary goal of being more broadly applicable. This goal is in line with other National Science Foundation (NSF) goals such as the EarthCube project led by the National Center for Atmospheric Research (NCAR) [16].

As a reference for building our model, a review paper by Meier (1991) discusses many models of the time and presents a plot of the "Altitude of Unit Optical Depth" ([17], Figure 2). Our initial aim was to match these results by combining the NRL density model and the MPI-Mainz UV-VIS database, providing a platform that can be adapted to fit the user's needs.

The initial inspiration for this work was to understand atmospheric absorption during sounding rocket flights, for example, the upcoming Full-sun Ultraviolet Rocket Spectrograph (FURST). FURST was motivated by a distinct lack of complete Vacuum Ultraviolet (VUV, 120-181 nm) spectra at high-resolution ($\mathcal{R} > 2 \cdot 10^4$, $\Delta V < 15$ km/s) of the full solar disk. A good example of the currently available data is SUMER, which has a good resolution ($\mathcal{R} \approx 7 - 16 \cdot 10^4$ from 67-161 nm) but did not capture the full disk (1" x 300" slit) [18–20]. SUMER can perform a raster across the surface, but it takes 31 hours to do so, limiting the temporal resolution of such a measurement. This and other slit-based spectrometers focus on narrow wavelength bands and cannot capture the full disk of the Sun over small time scales. FURST will provide not only an opportunity to observe the full disk of the Sun in such an important and wide wavelength range but should observe UV absorption from Earth's upper atmosphere.

In addition, we intend to launch a set of instruments from higher latitudes. The Hi-C Flare campaign will require waiting on the launch pad for hours in order to launch at the opportune time to capture an active flaring region on the Sun [21]. With this constraint, Poker Flat, Alaska is required to be used. However, at this latitude and extended launch windows, some atmospheric absorption will be inevitable. Predicting this effect will assist our team in adjusting the launch windows (time of year/day) for this exciting campaign.

In this paper, we discuss the development of an atmospheric absorption model that can be used for various purposes. To test one application, we simulate the potential impact on an instrument and utilize specific absorption features as fiducial points to aid in spectral calibration. Additionally, given a high spectral and radiometric resolution, we have the capability to analyze spectral images of past sounding rockets to uncover the fundamental physical properties of absorption-causing elements. *In-situ* measurements of the kind and at these altitudes can only

Journal of Physics: Conference Series **2544** (2023) 012001 doi:10.1088/1742-6596/2544/1/012001

be done with sounding rockets.

In Section 2, we introduce the methods for building up the spherical shells model, including refraction and atmospheric properties. We set up the model by introducing the optical depth calculation for a curved atmosphere with refraction. The model is then simplified for computation with imported databases for atmospheric density and molecular absorption cross-section. We finish the model by simulating a spectral image on our sounding rocket instrument with these absorption effects. Finally, we discuss two possible results from such an approach: spectral calibration and inversion of atmospheric information from raw data. In Section 3 we will summarize these results, and in Section 4 we discuss the potential uses for this technique and future work.

2. Methods

2.1. Differential Path Length

To introduce the necessary components for building up our model, we start with the optical depth. This term characterizes the amount of material that light will have to pass through before arriving at the detector. One normally uses Beer-Lambert's Law to calculate the optical depth τ, which requires knowledge of the absorption cross-section σ (cm^2/molecule) and the number density η (molecules/cm^3) as follows:

$$I = I_0 e^{-\tau},\ \tau(r,\lambda) = \int_r^\infty \sigma(r,\lambda)\,\eta(r')\,\frac{dr'}{\mu} \implies \tau_{i,j} = \sum_{i'=i}^{\infty} \sigma_{i',j}\eta_{i'}\Delta L_i, \tag{1}$$

where I is the intensity at the detector, I_0 is the intensity of the source. The optical depth varies with height r (km) and wavelength λ (nm). Since we are developing a database-driven model, we switch to index notation and label them as i and j, respectively. We write the transmission factor $T = I/I_0$ to measure how much light would be absorbed in the specified altitudes and wavelengths.

It is the convention to denote the zenith angle θ in the form $\mu \equiv \cos\theta$. In traditional models, the factor μ alone can account for the change to optical depth due to changing zenith angle (see left diagram in Figure 1). This is valid for a small angle θ, however, we want a more robust model that takes into account the curvature of the Earth's atmosphere, refraction, and changing altitude. With this in mind, we have combined it with the model altitude resolution Δr and will refer to it as the "differential path length," $\Delta L_i \equiv \Delta r_i/\mu_i$.

Solving for the differential path length is a primary contribution of this model. The diagram for setting up the geometrical argument is shown in Figure 1. To add further complexity, we have considered the cross-section σ to be a function of temperature. This dependence becomes very important at the altitude of unit optical depth, as our later results will highlight. Temperature is of course dependent upon altitude, so we write $\sigma \equiv \sigma(r,\lambda)$. We also attempt to account for the impact of refraction to ensure we have covered the most significant factors.

Using these given variables, we can calculate the adjusted ΔL as described in Figure 1. In this Figure, the spherical shells are shown on the left, along with 3 triangles that highlight the relevant trigonometry. Beginning at ground level, we solve for each ΔL along the path length from the observer to the source. The triangles here indicate the necessary angles and side lengths needed to solve for each ΔL_i. As in Equation 1, these non-uniform differential path lengths are what distinguish this setup from the simple model with uniform $\Delta L = \Delta r/\mu$.

There are many ways to solve these triangles for ΔL_i, but we have done so in a way that requires the intermediate angle θ_{Li} in order to prepare for the application of refraction. To solve for the differential path length ΔL_i, we employ simple trigonometric relations (see Appendix A) to work out that $\Delta L_0 = \sqrt{(r_0+\Delta r_0)^2 - r_0^2\left(1-\mu_0^2\right)} - r_0\mu_0$. This is the solution for the

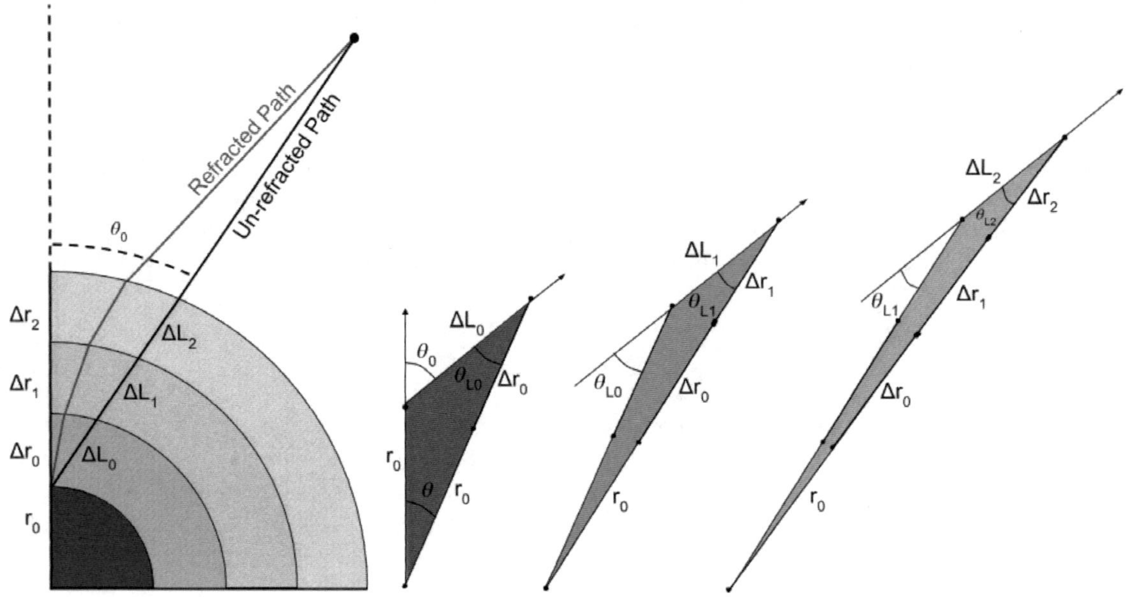

Figure 1. Cartoon of Geometry and Trigonometry to solve (not to scale). The spherical shells have an initial zenith angle that depends upon the starting point of the observation, r_0. The differential path length ΔL_i is labeled for a refraction/refraction-less setup. The next 3 diagrams show the relevant geometries at this initial height, with the effect of refraction exaggerated. The first triangle shows the necessary angles and side lengths in order to solve for ΔL_0. The next two show the progression in solving for ΔL_i along each additional shell. For the proper optical depth calculation in Equation 1, we must calculate all intermediate ΔL_i along the optical path at each subsequent height r_i.

first spherical shell at the initial altitude r_0. Extending to the generalized solution for each subsequent shell i in the differential path length ΔL_i:

$$\Delta L_i = \sqrt{\left(r_0 + \sum_{i'=0}^{i} \Delta r_{i'}\right)^2 - \left(r_0 + \sum_{i'=0}^{i-1} \Delta r_{i'}\right)^2 (1-\mu_i^2)} - \left(r_0 + \sum_{i'=0}^{i-1} \Delta r_{i'}\right)\mu_i, \tag{2}$$

$$\mu_i^2 = 1 - \left(\frac{r_0 + \sum_{i'=0}^{i-1} \Delta r_{i'}}{r_0 + \sum_{i'=0}^{i} \Delta r_{i'}}\right)^2 (1-\mu_{i-1}^2), \tag{3}$$

giving us a solution for the length of each discrete path between all shells. This solution is fed into Equation 1 to yield the total absorption at each particular wavelength for the starting altitude r_0. We can then adjust the starting altitude r_0 and perform the calculations again to build up the model for the absorption profile.

As mentioned in Section 1, a popular paper that describes atmospheric absorption in these wavelengths is [17]. We set the date, time, and location listed in Table 6 from that paper: March 21, 1980, 10:00 AM local time, at the GPS coordinates for White Sands Missile Range (WSMR), NM. From those details, we can query various databases and online resources for the initial zenith angle $\mu_0 \approx 0.7$ (from PySolar [22]), and the altitude and sea level for $r_0 \approx 6379.4$ km (from the USGS [23]). We also load into the program the PyNRLMSISE-00 atmospheric model data at a defined resolution $\Delta r = 1$ km [24]. Higher spatial resolutions can of course be specified as desired, but show no appreciable difference to the results presented in this paper.

In the calculations above, θ_{Li} is solved at each step and depends on the normal angle from the previous shell. This was done intentionally to provide a natural progression to refraction correction.

2.2. Refraction

We refer to [25] for a rigorous derivation, showing that refraction corrections are small for altitudes above 20 km and do not exceed 0.2°. For completeness, we have developed the tools for applying refraction to the intermediate angle θ_{Li} at the shell boundaries. For the applications in this paper, it is not necessary to apply these corrections Refer to Appendix A for more details.

From Snell's law, there exists an equivalence on either side of the shell "wall," namely

$$\mu_i^2 = 1 - \left(\frac{n_{i-1}}{n_i}\right)^2 \left(1 - \mu_{i-1}^2\right), \tag{4}$$

where n is the index of refraction of O2 and θ is the angle to the normal of the shell boundary. Substitution into Equation 3 yields a correction factor onto μ_i. If the density is low ($\rho \leq 1$ g/cm^{-3}), it can be expected that the refractive index n is near unity [26; 27]. From the Lorenz-Lorentz relation [28], the refraction index can be expressed relative to the previous refractive index as

$$n_i = \frac{\eta_i}{\eta_{i-1}} \left(n_{i-1} - 1\right) + 1, \tag{5}$$

where η is the number density and n is the refractive index. Using the same low-density assumptions, it is known through the Cauchy formula [29] and [25] that the dispersion formula for the refractive index as a function of wavelength can be expressed as

$$n_0 = A\left(1 + \frac{B}{\lambda^2}\right) + 1, \tag{6}$$

where constants A and B are tabulated from observations (see Appendix B).

A final complication related to refraction correction arises when we consider the initial zenith angle compared with the "perceived" zenith angle. As shown in Figure 1, the refracted path as perceived by the observer has a given angle θ_0 that is slightly higher than the actual zenith angle. As such, we must correct the given initial zenith angle such that the curved refraction path aligns with the location of the Sun. Since the correction for refraction seems to be small for our applications in this paper, it is left for future work. In the next section, we will put all of our functions together and remove the recursive relationships.

2.3. Assembling the Model

In order to avoid computationally expensive functions, we simplify our summations and nested expressions. Since we are using a uniform atmospheric density model, apply a uniform Δr. Expanding the nested functions is rather trivial and results in a much cleaner and faster computation (see Appendix B for complete derivation).

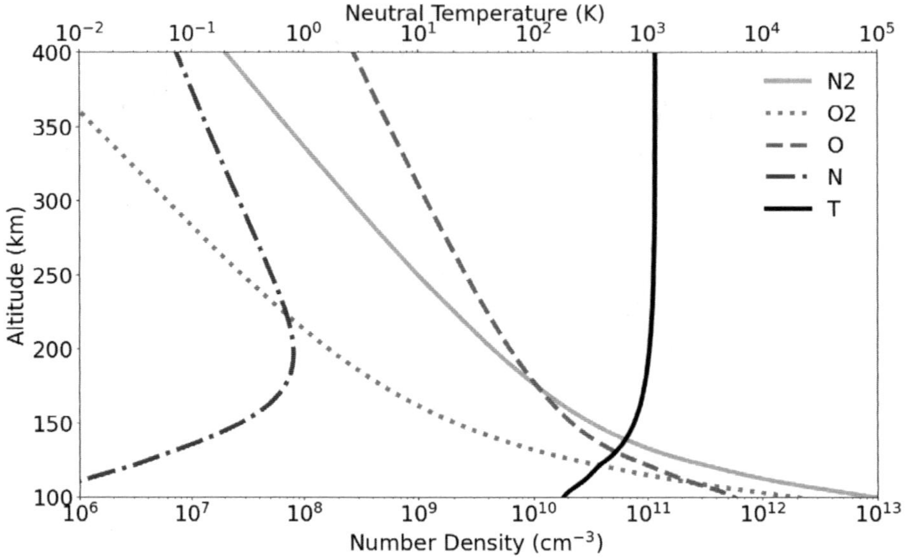

Figure 2. Densities of H, He, O, O_2, N, N_2, and Ar. These data show good agreement with [17], Figure 30.

$$\Delta L_i = \sqrt{\left[r_0 + (i+1)\,\Delta r_0\right]^2 - (r_0 + i\Delta r_0)^2\left(1-\mu_i^2\right)} - (r_0 + i\Delta r_0)\,\mu_i, \tag{7}$$

$$\mu_i^2 = 1 - \left(\frac{r_0 + \Delta r_0}{r_0 + (i+1)\,\Delta r_0}\right)^2 \left(\frac{n_0}{n_i}\right)^2 \left(1-\mu_0^2\right), \tag{8}$$

$$n_i = \frac{\eta_i}{\eta_0}\,(n_0 - 1) + 1. \tag{9}$$

This equation holds true for any $i \geq 0$. To avoid confusion with the symbols, recall again that the refractive index is n and the number density is η. Among the initial conditions needed are the initial altitude r_0 and the cosine of the zenith angle μ_0, which are given based on the location and time of day. As mentioned in Section 2.2, the refractive index n_0 should be a function of wavelength. This correction is ignored for now, and in fact refraction, in general, does not contribute to the results presented in Section 3 and can be set to a constant of unity. In the next sections, we discuss the models for number density η and the absorption cross-section σ.

2.4. Atmospheric Properties

In order to compute the expected absorption from the atmosphere, we need a model of the density of the material causing the absorption. We could begin by setting up the number density to follow the general exponential trend. However, a complex data-driven model has already been developed by the Naval Research Laboratory: the NRLMSISE-00 Model [2; 24]. This model provides users with the temperature and density for Hydrogen, Helium, Oxygen, O_2, Nitrogen, N_2, and Argon for specified dates and locations as a function of altitude.

The data plotted in Figure 2 is retrieved at the time of year and for the specified launch location from [17]. We can adjust the time of day as well to see the effect this has on the final model. Similarly, the zenith angle can be found in online tools such as the Solar Position

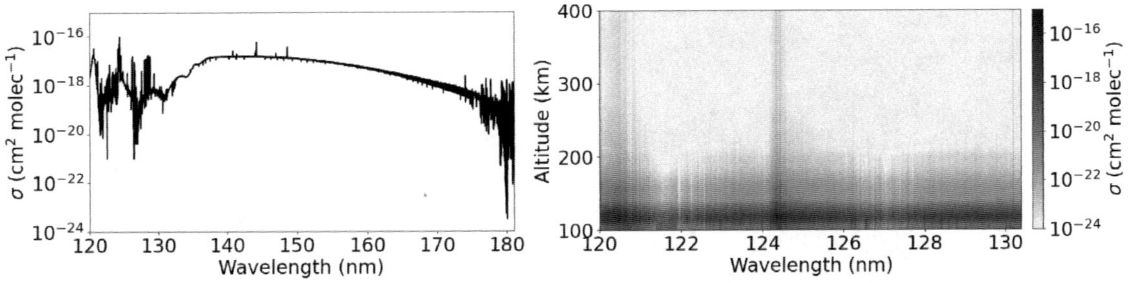

Figure 3. Absorption cross section for O_2 between 120-181 nm (left) and varying with Altitude between 120-130.4 nm. These data show agreement with [17], Figure 27.

Calculator provided by NOAA [30; 31] or the python integrated package "PySolar" [22]. For this model, at solar noon $\mu \approx 0.7$ at ground level.

The final piece of the optical depth calculation is the absorption cross-section. We show in Figure 3 the values for O_2 available from the MPI-Mainz UV-VIS Spectral Database [4–6] for the specified spectral range. Cross-sections are given in terms of cm^2 per molecule. This plot roughly matches the order of magnitude of [17], Figure 27, as well as a much older paper ([32], Figure 10).

The data set itself is based on laboratory measurements and detailed at various wavelengths and temperatures. As such, the data are sparse and some interpolation is needed to set up the continuous 2D array shown in Figure 3 (right). For this interpolation, we retain the given values while ensuring smooth variation between any missing points. This is the simplest method that preserves the general trend seen in the data. As noted later in Section 2.7, the lack of complete data is another motivation for detailing this model. It may be possible to fill in this missing information by looking at raw spectral data taken *in-situ* and determining the absorption cross-section from that measurement.

Note many interesting features here in Figure 3 (left). Firstly, the continuum is nicely visible, with a few small peaks near the middle part of the range. Given a large enough Signal-to-Noise Ratio (SNR), these stand-out peaks could aid in calibration. The highest value, nearing the -16 order at around 125 nm, will be useful as well. Finally, there are the Schumann-Runge bands at the far end, beyond 175 nm. This messy region could provide fiducials for spectral calibration [33].

Putting the cross-section data together with the number density is the final piece to calculating the optical depth as it varies with height and wavelength $\tau(r, \lambda)$. A useful representation of this result is the altitude of "unit optical depth." This is the location where $\tau = 1$ in the exponential term of Equation 1, and is shown as a function of altitude in Figure 4.

This result is in good agreement with [17], which also matched previous results from [34]. This plot makes sense when compared with the sharp drop in the density of O_2 in Figure 2. Additionally, the right plot in Figure 3 shows that this altitude has a significant shift in absorption cross-section. These effects combine to produce a sharp transition at around 80-120 km. The sharp peaks near the start of the range will certainly provide some calibration points, but the Schumann-Runge bands will most likely be too low for this specific test. In the next section, we simulate a set of noisy camera images for testing and analysis of this model.

2.5. Simulated Spectra

After calculating the optical depth, we can simulate an expected CCD signal by combining the transmission value T with the estimated photon flux I_0. In this section, we will explain

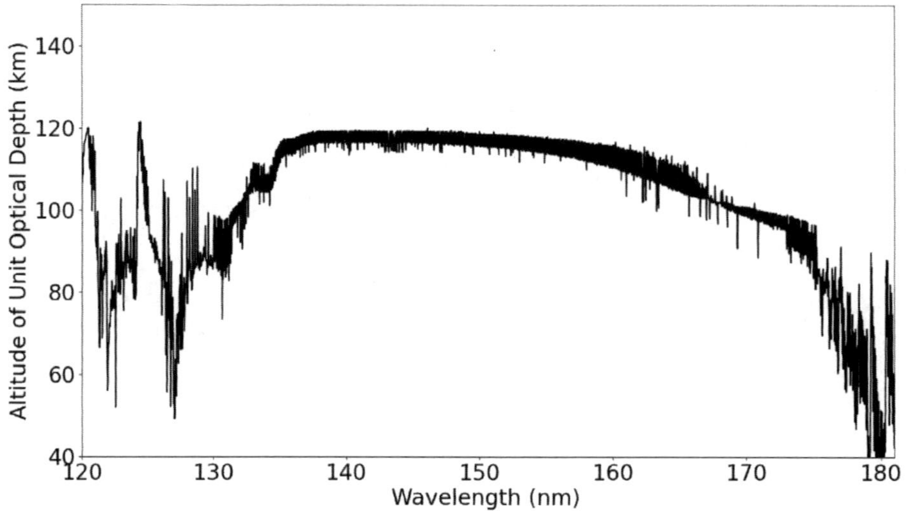

Figure 4. Altitude of Unit Optical Depth as calculated by Equation 1 and 7. Shows agreement with [17; 34]

the method for generating a simulated solar spectrum by using data from the High-Resolution Telescope and Spectrograph (HRTS) as a baseline for solar activity.

The HRTS spectrum is well known and is one of the better databases for the UV range [35]. Other sources exist such as AIA and LISARD, which present solar monitoring information for the general public [36]. As the name suggests, HRTS has a high spectral resolution (0.005 nm). This is necessary if we want to use it to predict what FURST will see, which is expected to have a similar or better resolution.

The Flare Irradiance Spectral Model (FISM2) is also used. Described in [37] and available on [36], this model puts together actual measurements alongside an empirical model of solar variability to produce a more complete data set of UV spectral irradiance at 0.1 nm resolution.

The intensity of the HRTS Quiet Sun (QS) signal is shown in Figure 5. Since the noise floor for HRTS is high, we also use the FISM2 model as a reference from the same date (March 1980). The HRTS spectra shown here are originally given in units of erg/s/sr/Å/cm^2. We convert to ph/s/nm/cm^2 by using the size of the full disk ($9.352 \cdot 10^{-3}$ sr), but an additional factor of $\approx$ 1.4x is used in Figure 5 to scale the intensity to match the FISM2 model.

This difference is relatively small but could be attributed to the scaling of HRTS to the full disk. Additionally, the QS spectra do not include additional irradiance from Explosive Events (EEs), Active Regions (ARs), etc. Even so, we prefer to use the HRTS spectra over the FISM2 model in this application because it is given at a much higher spectral resolution. The best way to obtain a full-disk high-resolution spectrum would be to measure it directly, which is one of the primary motivations for developing an instrument like FURST.

To simulate the spectral signals we need to fold both measurements into the instrument functions for FURST. This conversion from the raw-data values to Data Numbers (DNs) is done by having knowledge of the instrument, cameras, filters, and other effects such as noise, and can be found in the Appendix C. Of note, there will be a spectral plate scale of $\approx$ 0.005 nm per pixel. This defines the 1-dimensional linear wavelength-to-pixel mapping function used in this paper to demonstrate the model's use for calibration.

Additionally, specifying an exposure time requires us to implement the expected flight profile, since a single exposure occurs at various altitudes. For this setup, we have a 10-second exposure

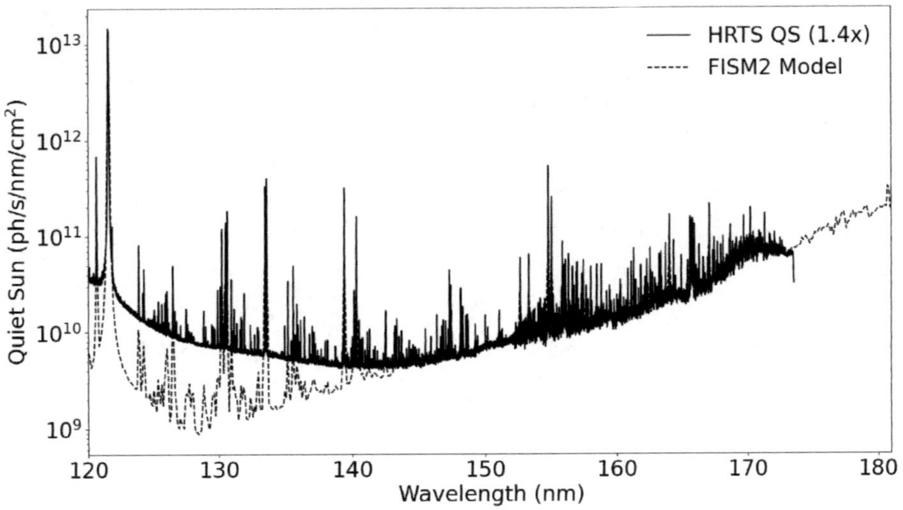

Figure 5. The solid line shows the HRTS spectra for the Quiet Sun (QS) scaled by 1.4x to match the FISM2 model shown by the dashed line.

window at the apogee near 254 km, where there is effectively no absorption, and another at the bottom end of the flight between 115-132 km, where there is much more absorption.

With these conversions in place, a noisy image is generated for a FURST camera signal. The top image in Figure 6 is the 2D simulated signal for the window between 115-132 km. The noise can be seen here, but the spectral features are clear in the bottom plot. The top solid line is the FURST signal averaged across all rows from the simulated 2D image. The red dashed line beneath it contains the FISM2 simulated signal as described in Figure 5.

An important comparison to this available data is the estimated FURST noise floor. To calculate this value, we take the various sources of noise from Table C1 in Appendix C and work out the highest resulting output signal from an input signal of zero photons. This works out to an output of approximately 2.5 DN shown as the bottom solid red line in Figure 6.

On average, the Standard Error of the Mean (SEM) for these signals is much less than 1% so it is not displayed here. It is also encouraging to see that both signals are many orders of magnitude higher than the estimated noise floor for FURST. This helps to point out the clear improvements being made with the instrumental capabilities FURST will provide.

With the appropriate conversions and noise accounted for, we now have a method for computing atmospheric absorption at the altitudes along a sounding rocket flight profile. This will allow us to see if the instruments have enough SNR and spectral resolution to be useful for calibration or inversion of atmospheric properties.

2.6. Calibration

Before a spectral image from an instrument can be properly analyzed for solar physics, it must first be calibrated. There are two different types of calibration of interest to us here: radiometric and spectral. In this context, radiometric calibration refers to determining the conversion of the signal at each pixel from the computer signal in terms of data number (DN) back into physical units (W cm^{-2} s^{-1} sr^{-1} $Å^{-1}$). Likewise, spectral calibration involves determining the fitting parameters for recasting the pixel numbers in terms of their known wavelength values. One common parameter is the spectral plate scale (in nm per pixel), although typically mapping functions are 2D nonlinear functions.

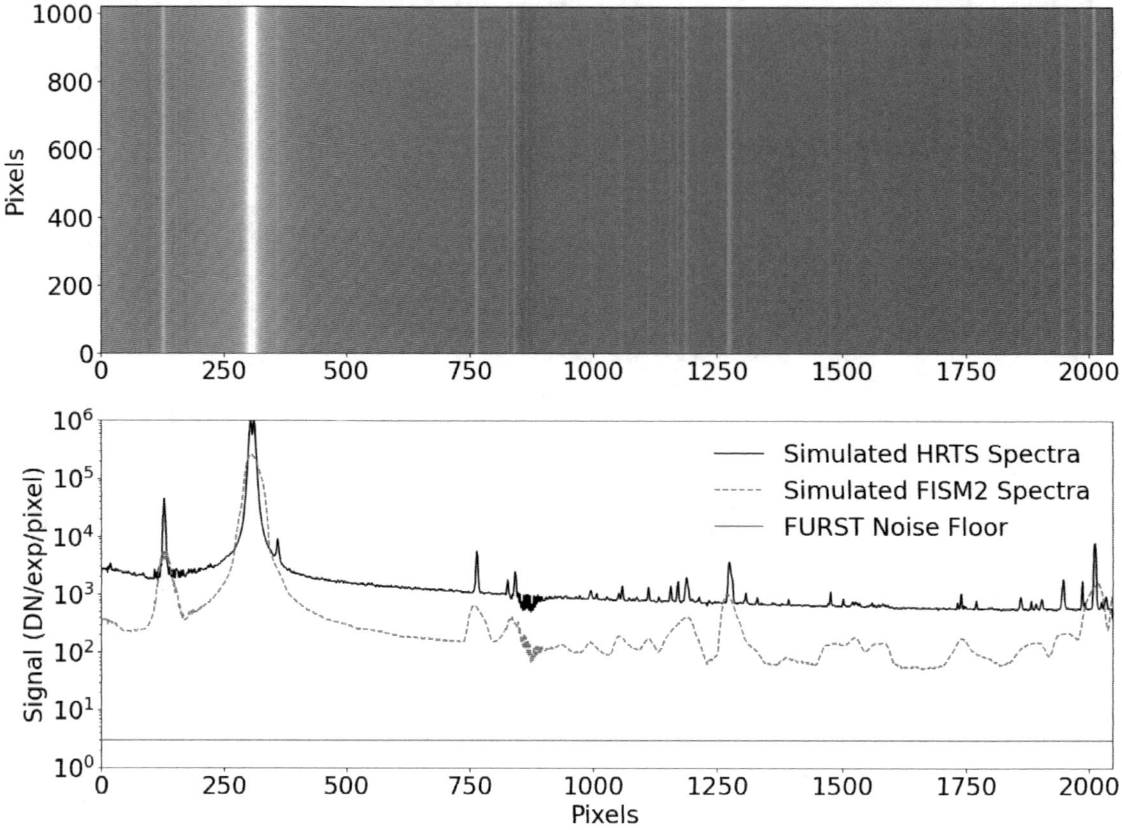

Figure 6. Top: A simulated 2D noisy image from a 10-second exposure between 115-132 km. Bottom: The median signal level across each row (upper/solid black line) with the estimated solar continuum (middle/red dashed line) and the estimated noise floor (lower/solid red line) for the FURST instrument.

To use the identified absorption features for spectral calibration, we use a spherical shells model to properly account for the levels of expected noise and test whether or not we have sufficient resolution to be able to fit these absorption features. In Figure 7, we show an example of this. Here we see the simulated signals from a higher altitude S_H and a lower altitude S_L. In the right plots, we compare the ratio of those signals with the absorption cross-section plot, using the corresponding altitude.

A simple Gaussian is used to map out corresponding pixel and wavelength pairs. From just two of these pairs, we recalculate the spectral plate scale as to within $\approx 0.5\%$ error. More sophisticated wavelength mapping programs are to be used in practice to develop our calibration beyond just solving for the linear parameter. This is a promising result that shows we indeed have sufficient SNR for a viable *in-situ* calibration, as well as that these features are ubiquitous across the spectrum.

2.7. Inversion

Looking back to the models used to generate the spectral images in the previous sections, the NRL density model is well-trusted but the absorption cross-section database is sparse. This motivated an investigation into a specific application for sounding rockets. It may be possible

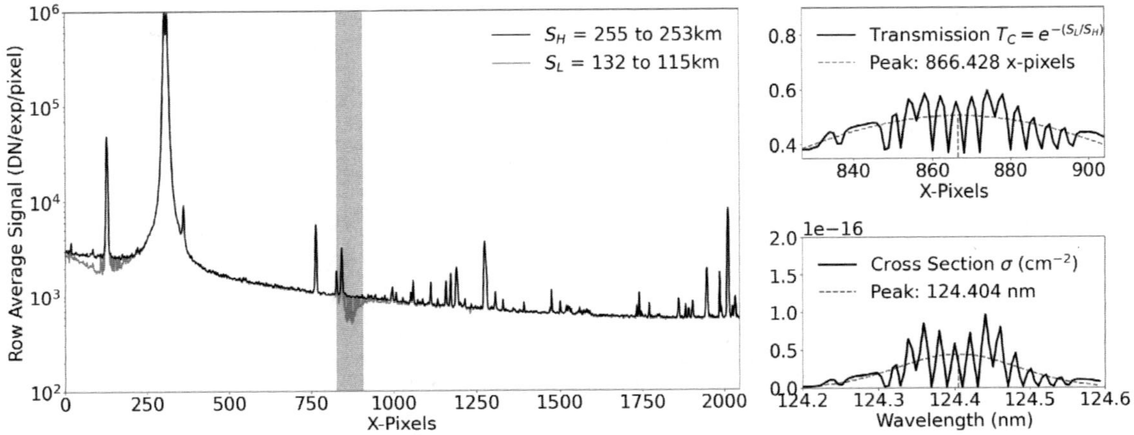

Figure 7. Simulated signal from 10-second exposures, row-averaged. The left plot shows the full wavelength range for one channel. The top right is the transmission factor of a zoomed-in region, calculated as shown from the ratio between the simulated data at the higher altitude (S_H) and the lower altitude (S_L). The lower right plot compares this segment with a known peak in absorption cross-section.

to invert the atmospheric absorption from spectral images taken during a flight.

From Equation 1, the density and absorption cross-section is folded into the integral that yields optical depth. If spectral images are taken over a relatively small range of altitudes, temperature variations will be small and the cross-section σ can be considered independent of height. Thus, the term can be separated following

$$T_{i,j} = e^{-\sigma_j \sum_{i'=i}^{\infty} \eta_{i',j} \Delta L_i} \implies \frac{-\ln T_{i,j}}{\sum_{i'=i}^{\infty} \eta_{i',j} \Delta L_i}, = \sigma_j, \tag{10}$$

where the transmission factor T must be found at altitude i and wavelength j. If a spectral image is taken near apogee then absorption is effectively negligible. Taking the ratio of an observation from a lower height with one taken at the maximum height will give a proxy of T at that lower altitude i. Then we can apply Equation 10 to invert the absorption cross-section using our model.

We can test this with the simulated results from FURST described in Section 2.6 and shown in Figure 7 (left). We take the simulated noisy spectral images taken at the corresponding highest ($\approx$ 254 km) and lowest ($\approx$ 115-132 km) altitudes. Aside from the noise, the main difference between these two spectral images can be attributed to the atmospheric absorption effect. We then divide the lower image from the upper images, as in Equation 10, and recover a "measured" optical depth.

The results of this test are shown in Section 3, Figure 9. As an alternative test, we could place more trust in the cross-section data and attempt to solve for a density profile. The results of this experiment are shown in Appendix D.

3. Results

We first show a comparison of the differential path length as calculated by the "simple" model (where $\Delta L = \Delta r/\mu$) with the result from the spherical-shells model (where ΔL is calculated as in Equation 7). An example is shown in Figure 8, generated at varying initial zenith angles (μ_0 = 0.2, 0.4, and 0.6) to highlight the effect. We can see the difference between the use of

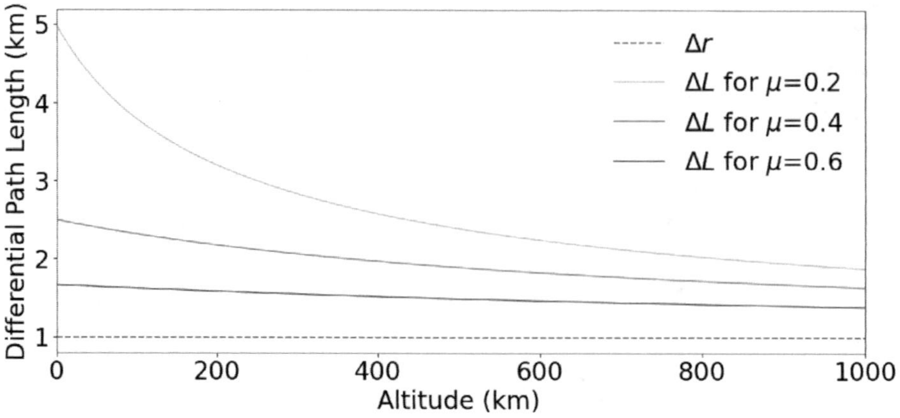

Figure 8. Comparison of the Differential Path Length ΔL at different initial zenith angles μ in shades of gray. The bottom red dashed line indicates the altitude resolution Δr of 1 km in this example. The simple model would calculate ΔL as $\Delta r/\mu$, which corresponds to the initial point on the left, at 0 km.

uniform and non-uniform differential path lengths from the simple model and the shells model, respectively.

The simple model would estimate a uniform differential path length as $\Delta r/\mu$, which corresponds to the initial points (at 0 km) in Figure 8. However, the spherical shells model allows for this to be a non-uniform value. At higher shells, the angle between the shell and the path approaches normal ($\mu \longrightarrow 1$), forcing the value of the differential path length to approach the shell spacing ($\Delta L \longrightarrow \Delta r$) (bottom/red dashed line). To help understand the effect of this, we can compare the sum of the ΔL values between each model. When $\mu \approx 0.2$, the shells model puts this value at about half of the simple model. As we approach a zenith angle greater than 0.6 the difference drops to less than 10%. This model is still in-line with the previous work by [17] since the time of day provided a high zenith angle. These results, however, predict slightly less absorption overall for our flight path than simpler models.

Next, as can be seen in Figures 6 and 7, once the signal is transmitted through the instrument we have enough SNR to tell where the absorption lines are. These fiducials provide calibration points to verify the pixel-to-wavelength mapping of the instrument to within 0.5%, even though we only considered 2 lines in this basic example. Furthermore, this calibration only considered one wavelength channel for FURST. By coupling more fiducials and a broader range we hope to add significantly to the wavelength calibration of the data through this *in-situ* method.

Finally, we solve for the absorption cross-section σ required to produce the spectra seen by an instrument like FURST. As described in Section 2.7, this inversion required taking the ratio of the spectral image taken at apogee with one from a lower height (see Figure 7) and is formulated in Equation 10.

In Figure 9, the black and red lines represent the known and the computed absorption cross-section σ, smoothed and scaled for ease of comparison. We clearly reproduce the values quite well, but it is also apparent where noise has made an impact. There are a few deep lines (around 122 nm and 127 nm) that are a result of this.

Additionally, the absolute values are a bit off. From our testing, this can be attributed to the temperature change at the altitudes where this simulated signal occurs There is a sharp change in the temperature, and thus the absorption cross-section values. To better reproduce the laboratory values, we can employ a more nuanced calculation. For example, we have utilized

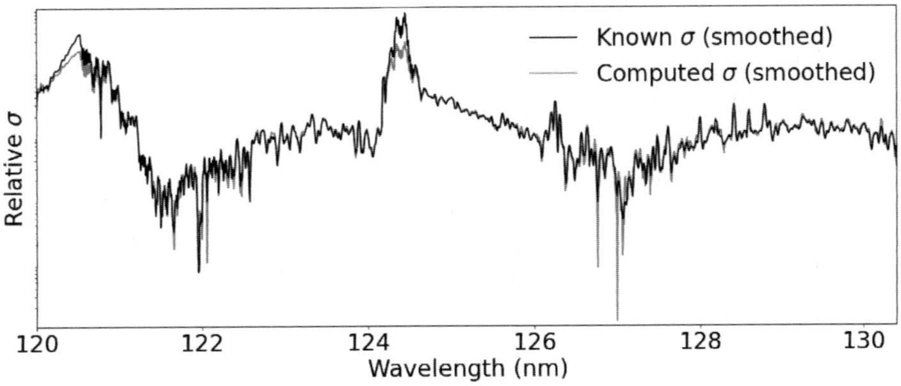

Figure 9. Recalculated absorption cross-section from simulated images (red) compared with the known data (black). Both are smoothed and scaled for easier comparison.

a weighted average based on the flight profile. However, more work must be done to understand what is missing.

Regardless, the primary features of interest are the peak locations rather than their absolute values, and this result shows that they are almost completely recovered. There is also a tendency for higher values to have less error. This is because there is less noise associated with the spectral image signal from which it is calculated. It may not be possible to perform this calculation from instruments with a low SNR. This further stresses the importance of radiometric calibration, and the added value such an effort brings.

This method shows a broadly applicable technique for the inversion of atmospheric properties from existing data. It may help significantly to fill gaps in the predicted absorption points in our databases. We have set up this model to look at molecular oxygen (O_2) since this is the primary contributor to UV absorption. However, other molecules can be explored with this model that may be of greater interest to atmospheric scientists (e.g. Ozone, Methane, Aerosols, etc.).

4. Discussion

This updated model has been shown to accurately reproduce the expected atmospheric absorption predicted by previous models using modern techniques. Our results match previous reports well while accounting for the curvature of the Earth and its effects on a non-uniform differential path length. For zenith angles above $\mu > 0.6$, a simple uniform differential path length will change the optical depth by a factor of less than 10%. This model is especially suited to applications with a low zenith angle, such as heliophysics tests performed outside of solar noon, at high latitudes, or with a low flight apogee. With knowledge of the proper absorption effects, this model can assist in extending the launch capabilities for such missions.

This model's biggest benefit for our team, though, is in improving our understanding of the UV absorption effect in FURST data products. It will also enable us to set limitations or requirements for future launch conditions. Missions with a desire to measure or avoid UV absorption can benefit from such understanding.

There are of course many improvements that can be made. Extrapolating the given density data set out above 1000 km, and increasing the resolution, could improve the resolved density profile or better predict the absorption amount. There also remain other atmospheric effects that we anticipate being useful for the calibration and inversion of the FURST data products. Hydrogen and Oxygen resonant absorption adds additional absorption features that can be utilized for calibration and cross-section inversions. A thermally expanded atmosphere due to solar sub-storms, or other such responses to flare events, will have the effect of increasing the altitude of unit optical depth [38].

The largest improvement to be made is with the sparseness of the absorption cross-section data. Around the upper heights of the rocket flight, the temperature does not vary all that much. However, the NRL model shows significant variation in temperature at the altitudes corresponding to the beginning and end of the shutter open time. We have attempted to use the available data and have adjusted our code to accept whatever new data becomes available. Rather than rely on laboratory measurements, we could look into calculating the theoretical spectral width of the O_2 absorption features. Not only should spectral widths change with temperature, but some lines may only appear at certain temperature ranges. Understanding the physics behind calculating the resonant absorption lines will likely lead to an understanding of how to solve this problem.

Additional work on FURST involves improved radiometric calculations for tracking errors throughout the calibration system and flight instrument and has been discussed in a separate paper [39]. With these high-resolution capabilities, the noise floor of the FURST instrument will be well below the expected solar continuum, making the absorption profile easier to resolve. Following our spherical shells model, *in-situ* wavelength calibration and radiometric inversion of the atmospheric absorption cross-section should be possible.

5. Acknowledgements

The primary author wishes to thank their advisors, mentors, and colleagues that helped with countless meetings, advice, and support along the way: Amy, Charles, Larry, Gen, Adam, Gary, Michael, Keyvan, Clayton, Juan, all of the staff at UAH for the funding support, his wife Crystal, but most importantly, God (Matthew 6:33-34). This material is based upon work supported by the NSF EPSCoR RII-Track-1.2a (Non-invasive plasma diagnostics for LTP) Cooperative Agreement OIA-1655280. Any opinions, findings, conclusions, or recommendations expressed in this material are those of the authors and do not necessarily reflect the views of the National Science Foundation.

References

[1] Friedman H, Lichtman S W and Byram E T 1951 *Phys. Rev.* **83**(5) 1025–1030 URL `link.aps.org/doi/10.1103/PhysRev.83.1025`

[2] Moussa M and Kuznetsova M 2022 Nrlmsis atmosphere model Tech. Rep. Version 1.0 and 2.0 NASA Community Coordinated Model Center `kauai.ccmc.gsfc.nasa.gov/instantrun/msis`

[3] Gordon I E, Rothman L S, Hill C, Kochanov R V, Tan Y, Bernath P F, Birk M, Boudon V, Campargue A, Chance K *et al.* 2017 *Journal of Quantitative Spectroscopy and Radiative Transfer* **203** 3–69

[4] Keller-Rudek H, Moortgat G K, Sander R and Sörensen R 2014 The mpi-mainz uv/vis spectral atlas of gaseous molecules of atmospheric interest Tech. rep. Earth System Science Data `uv-vis-spectral-atlas-mainz.org/uvvis/cross_sections/Oxygen/O2.spc`

[5] Keller-Rudek H, Moortgat G, Sander R and Sörensen R 2013 *Earth System Science Data* **5** 365–373

[6] Sander R, Keller-Rudek H, Moortgat G and Sörensen R 2014 *EGU General Assembly Conference Abstracts* p 1753

[7] Smette A, Sana H, Noll S, Horst H, Kausch W, Kimeswenger S, Barden M, Szyszka C, Jones A, Gallenne A *et al.* 2015 *Astronomy & Astrophysics* **576** A77

[8] Kausch W, Noll S, Smette A, Kimeswenger S, Barden M, Szyszka C, Jones A, Sana H, Horst H and Kerber F 2015 *Astronomy & Astrophysics* **576** A78

[9] Jones A, Noll S, Kausch W, Szyszka C and Kimeswenger S 2013 *Astronomy & Astrophysics* **560** A91

[10] Noll S, Kausch W, Barden M, Jones A, Szyszka C, Kimeswenger S and Vinther J 2012 *Astronomy & Astrophysics* **543** A92

[11] Lean J L 1980 *Atmospheric ultraviolet absorption spectroscopy* Ph.D. thesis University of Adelaide, Dept. of Physics

[12] He W, Wu K, Feng Y, Fu D, Chen Z and Li F 2019 *Remote Sensing* **11** 2702

[13] Sun K, Gordon I E, Sioris C E, Liu X, Chance K and Wofsy S C 2018 *Geophysical Research Letters* **45** 5779–5787

[14] Sun K, Yousefi M, Chan Miller C, Chance K, González Abad G, Gordon I E, Liu X, O'Sullivan E, Sioris C E and Wofsy S C 2022 *Atmospheric Measurement Techniques Discussions* 1–37

[15] Degenstein D A 2000 *Atmospheric volume emission tomography from a satellite platform.* (The University of Saskatchewan)

[16] Abernathey R *et al.* 2017 *GSTDTAP*

[17] Meier R 1991 *Space Science Reviews* **58** 1–185

[18] Curdt W, Brekke P, Feldman U, Wilhelm K, Dwivedi B, Schühle U and Lemaire P 2001 *Astronomy & Astrophysics* **375** 591–613

[19] Peter H 1999 *The Astrophysical Journal* **516** 490

[20] Peter H and Judge P 1999 *The Astrophysical Journal* **522** 1148

[21] Savage S, Winebarger A, Glesener L, Golub L, Chamberlin P, Hi-C Flare Team, Foxsi-4 Team and Snifs Team 2021 *American Astronomical Society Meeting Abstracts* (*American Astronomical Society Meeting Abstracts* vol 53) p 313.15

[22] Stafford B 2018 pysolar URL `doi.org/10.5281/zenodo.1461066`

[23] 2021 URL `nationalmap.gov/epqs/`

[24] Bender S 2022 Pynrlmsise-00 URL `pynrlmsise00.readthedocs.io`

[25] Born M and Wolf E 2013 *Principles of optics: electromagnetic theory of propagation, interference and diffraction of light* (Elsevier)

[26] Liu Y and Daum P H 2008 *Journal of Aerosol Science* **39** 974–986

[27] Phillips P 1920 *Proceedings of the Royal Society of London* Series A, Containing Papers of a Mathematical and Physical Character **97** 225–240

[28] Buckingham A D and Graham C 1974 *Proceedings of the Royal Society of London* A. Mathematical and Physical Sciences **337** 275–291

[29] CAUCHY L 1830 *bull. des. sc. maht.* **14** 9 URL `cir.nii.ac.jp/crid/1570572699363130880`

[30] Cornwall C, Horiuchi A and Lehman C 2022 Noaa esrl solar position calculator Tech. rep. NOAA Earth System Research Laboratories Global Monitoring Laboratory `gml.noaa.gov/grad/solcalc/azel.html`

[31] NOAA 2022 Noaa esrl solar calculator Tech. rep. NOAA Earth System Research Laboratories Global Monitoring Laboratory `gml.noaa.gov/grad/solcalc/`

[32] Watanabe K, Zelikoff M and Inn E C 1953 Absorption coefficients of several atmospheric gases Tech. rep. Air Force Cambridge Research Labs Hansom AFB MA

[33] Yoshino K, Freeman D and Parkinson W H 1984 *Journal of physical and chemical reference data* **13** 207–227

[34] Lean J 1987 *Journal of Geophysical Research: Atmospheres* **92** 839–868 (*Preprint* `agupubs.onlinelibrary.wiley.com/doi/pdf/10.1029/JD092iD01p00839`) URL `agupubs.onlinelibrary.wiley.com/doi/abs/10.1029/JD092iD01p00839`

[35] Brekke P 1993 *The Astrophysical Journal Supplement Series* **87** 443–450

[36] of the University of Colorado-Boulder R 2014 Lasp interactive solar irradiance datacenter (lisard) Tech. rep. Laboratory for Atmospheric and Space Physics, University of Colorado-Boulder `lasp.colorado.edu/lisird4`

[37] Chamberlin P C, Eparvier F G, Knoer V, Leise H, Pankratz A, Snow M, Templeman B, Thiemann E M B, Woodraska D L and Woods T N 2020 *Space Weather* **18** e2020SW002588 e2020SW002588 10.1029/2020SW002588 (*Preprint* `agupubs.onlinelibrary.wiley.com/doi/pdf/10.1029/2020SW002588`) URL `agupubs.onlinelibrary.wiley.com/doi/abs/10.1029/2020SW002588`

[38] Jacchia L G and Slowey J 1966 *SAO Special Report* **207**

[39] Vigil G D, Winebarger A, Rachmeler L, Donders N, Athiray P S, Kobayashi K and Kankelborg C 2021 *Journal of Astronomical Telescopes, Instruments, and Systems* **7** 035009

Appendix A. Details of Differential Path Length Formula

The following shows more details in the derivation of the differential path length calculations. Refer to Figure 1 as needed. To solve for the differential path length ΔL_i, we first employ the Law of Sines. Firstly, since $\sin(\pi - \theta_0) = \sin\theta_0$, then

$$\frac{\sin\theta_0}{r_0 + \Delta r_0} = \frac{\sin(\theta_{L0})}{r_0} \implies \theta_{L0} = \sin^{-1}\left(\frac{r_0}{r_0 + \Delta r_0}\sin\theta_0\right). \quad (A.1)$$

Then we use the 180° rule to solve for $\theta = \theta_0 - \theta_{L0}$. Finally, after some algebra, we use the Law of Sines again to solve for ΔL_0.

$$\frac{\sin\theta_0}{r_0 + \Delta r_0} = \frac{\sin\theta}{\Delta L_0} \implies \Delta L_0 = \sqrt{(r_0 + \Delta r_0)^2 - r_0^2\left(1 - \mu_0^2\right)} - r_0\mu_0, \quad (A.2)$$

where we allow the convention of $\mu_0 = \cos\theta_0$. Extending to each subsequent shell, we can see the pattern for ΔL_i as

$$\Delta L_0 = \sqrt{(r_0 + \Delta r_0)^2 - r_0^2\left(1 - \mu_0^2\right)} - r_0\mu_0 \quad (A.3)$$

$$\Delta L_1 = \sqrt{(r_0 + \Delta r_0 + \Delta r_1)^2 - (r_0 + \Delta r_0)^2\left(1 - \mu_1^2\right)} - (r_0 + \Delta r_0)\,\mu_1 \quad (A.4)$$

$$\Delta L_2 = \sqrt{(r_0 + \Delta r_0 + \Delta r_1 + \Delta r_2)^2 - (r_0 + \Delta r_0 + \Delta r_1)^2\left(1 - \mu_2^2\right)} - (r_0 + \Delta r_0 + \Delta r_1)\,\mu_2 \quad (A.5)$$

$$\vdots$$

$$\Delta L_i = \sqrt{\left(r_0 + \sum_{i'=0}^{i}\Delta r_{i'}\right)^2 - \left(r_0 + \sum_{i'=0}^{i-1}\Delta r_{i'}\right)^2\left(1 - \mu_i^2\right)} - \left(r_0 + \sum_{i'=0}^{i-1}\Delta r_{i'}\right)\mu_i, \quad (A.6)$$

$$\mu_i = \sqrt{1 - \left(\frac{r_0 + \sum_{i'=0}^{i-1}\Delta r_{i'}}{r_0 + \sum_{i'=0}^{i}\Delta r_{i'}}\right)^2\left(1 - \mu_{i-1}^2\right)}, \; i > 0, \quad (A.7)$$

$$\mu_0 = \cos\theta_0. \quad (A.8)$$

From Snell's law of refraction, there exists an equivalence on either side of the shell "wall," namely

$$n_i \sin\theta_i = n_{i-1}\sin\theta_{i-1} \implies \mu_i^2 = 1 - \left(\frac{n_{i-1}}{n_i}\right)^2 \left(1 - \mu_{i-1}^2\right), \tag{A.9}$$

where n is the index of refraction of O_2 and θ is the angle to the normal of the shell boundary. Upon substitution into Equation A.7, we end up with a correction factor onto μ_i given by the ratio of the index of refraction at each subsequent shell (n_{i-1}/n_i). The relationship between density and refractive index can be expressed linearly if $\rho \leq 1$ g/cm^{-3} (see [26; 27], and others). For O_2, this corresponds to a number density $\eta_{O2} \leq 10^{22}$ atoms/cm^3). If the density is low, it can be expected that the refractive index $n \to 1$. From the Lorenz-Lorentz relation [28] we can calculate this as

$$\frac{n^2-1}{n^2+2}\rho^{-1} = \frac{N_A\alpha}{3\epsilon_0} \overset{n\to 1}{\Longrightarrow} n = \frac{N\alpha_0}{2\epsilon_0}\rho + 1. \implies n_i = \frac{\eta_i}{\eta_{i-1}}(n_{i-1}-1)+1. \tag{A.10}$$

In Equation A.10, N_A and ϵ_0 are constants. The polarizability α is highly dependant upon density and wavelength. However, under linearity approximation at low density, the dependency can be expressed as a function of number density η. Taking the ratio of refractive indices of subsequent shells removes those constants, and only a single value for the refractive index at the instrument starting location n_0 is needed.

Using those same low-density assumptions, it is known through the Cauchy formula [29] and [25] that the dispersion formula for the refractive index as a function of wavelength can be expressed as

$$n_0 = A\left(1 + \frac{B}{\lambda^2}\right) + 1, \tag{A.11}$$

where constants A and B are tabulated from laboratory measurements to be $2.663 \cdot 10^{-4}$ and $5.07 \cdot 10^3$ nm^2 respectively. A more precise dispersion relation can be formulated using polarizability functions, but this is not necessary for a first-order solution. Since we have shown that the refractive index can be expressed as a function of wavelength, this implies that the differential path length is as well.

Appendix B. Simplifying Functions

There is a fair amount of simplification that occurs when we see how the nested functions begin to cancel out terms. For the zenith angle, the recursive functions terminate at $i = 0$ as follows.

$$\mu_i^2 = 1 - \left(\frac{r_0 + \sum_{i'=0}^{i-1} \Delta r_{i'}}{r_0 + \sum_{i'=0}^{i} \Delta r_{i'}}\right)^2 \left(\frac{n_{i-1}}{n_i}\right)^2 (1 - \mu_{i-1}^2) \tag{B.1}$$

$$= 1 - \left(\frac{r_0 + \sum_{i'=0}^{i-1} \Delta r_{i'}}{r_0 + \sum_{i'=0}^{i} \Delta r_{i'}}\right)^2 \left(\frac{n_{i-1}}{n_i}\right)^2 \left(\frac{r_0 + \sum_{i'=0}^{i-2} \Delta r_{i'}}{r_0 + \sum_{i'=0}^{i-1} \Delta r_{i'}}\right)^2 \left(\frac{n_{i-2}}{n_{i-1}}\right)^2 (1 - \mu_{i-2}^2) \tag{B.2}$$

$$= 1 - \left(\frac{r_0 + \sum_{i'=0}^{i-1} \Delta r_{i'}}{r_0 + \sum_{i'=0}^{i} \Delta r_{i'}}\right)^2 \left(\frac{r_0 + \sum_{i'=0}^{i-2} \Delta r_{i'}}{r_0 + \sum_{i'=0}^{i-1} \Delta r_{i'}}\right)^2 \cdots \left(\frac{r_0 + \Delta r_0}{r_0 + \Delta r_0 + \Delta r_1}\right)^2 \left(\frac{n_{i-1}}{n_i}\right)^2 \left(\frac{n_{i-2}}{n_{i-1}}\right)^2 \cdots \left(\frac{n_0}{n_1}\right)^2 (1 - \mu_0^2) \tag{B.3}$$

$$= 1 - \left(\frac{r_0 + \Delta r_0}{r_0 + \sum_{i'=i}^{i} \Delta r_{i'}}\right)^2 \left(\frac{n_0}{n_i}\right)^2 (1 - \mu_0^2)\,. \tag{B.4}$$

Likewise, the formula for refractive index reduces to Equations 9. To obtain the final expression for ΔL, we invoke the uniform shell-spacing $\Delta r_i = \Delta r$. Thus, summations $\sum_{i'=0}^{i} \Delta r_i$ become $(i+1)\Delta r$. This returns Equations 7 and 8.

Appendix C. Conversion of Signal to Data Numbers

A spectral signal is often given in units of ergs/s/Å/sr/cm^2. Table C1 shows the conversion process to units Data Numbers per exposure per pixel. Firstly, we convert the energy from ergs to photons using the photon energy. Then we have to account for the spectral plate scale for each wavelength channel in FURST, which is $\approx$ 10.4 nm across a width of 2048 pixels. Since we are observing the full disk of the sun ($\approx$32 arc-minutes), then its size in our field of view will be roughly constant at the value shown.

The signal is then attenuated by the instrument optics and the physical size of the CCD. Finally, the photon count rate is spread across the height of the CCD, which if the optics are aligned properly should match the number of rows of pixels. In folding the photon signal into the instrument response function, we take into account the charge spreading of the signal across adjacent pixels. This width σ is given from the FURST design and is around a pixel in size depending on the wavelength. We compute the line-spread function using a standard Gaussian, $I_0 \cdot \exp(-(\lambda - \lambda_0)/2\sigma^2)$, where λ_0 is taken as the HRTS signal mapped onto the defined λ space.

The camera sends this signal along the readout channels after a set exposure time, which for this testing we have set to 10 seconds. In order to count up the total number of photons during each exposure, we use the flight profile to fold in the hanging transmission factor, as calculated by Equation 1 and shown in the Results in Figure 4. In either case, it is at this point that the signal is saved as an integer number of photons per exposure per pixel.

Variable	Source	Values	Units
HRTS Signal	Raw Data	$\approx 414^{+11912}_{-307}$	erg/s/Å/sr/cm^2
Photon Energy	$E_{ph} = h \cdot c \cdot \lambda^{-1}$	$\approx 1.32 \cdot 10^{-11}$	erg/ph
Spectral Plate Scale	Optical Setup	$5.078 \cdot 10^{-3}$	Å/x-pixel
Solar Disk Area	Observation	$9.352 \cdot 10^{-3}$	sr
Effective Area	GA · Refl. · Trans. · QE	$\approx 4.176 \cdot 10^{-5}$	cm^2
Number of Rows	Variable	1024	y-pixel
Exposure Time	Variable	10	s/exp
Atmospheric Absorption	Model	0.0-1.0	unit-less
Photon Signal	Converted	$\approx 2070^{+91130}_{-1740}$	ph/exp/pixel
Photon Noise	Poisson	$\approx \pm 110.5\ (3\sigma)$	ph/pixel
Electrons per Photon	E_{ph}/E_{eh}	≈ 2.26	e$^-$/ph
Electron Noise	Readout, Fano, Shot, etc.	$\approx \pm 25\ (3\sigma)$	e$^-$/pixel
Gain	Variable	1.0	DN/e$^-$/pixel
Gain Bias	Variable	0 (will be 3000)	DN/pixel
Camera Signal	Converted	$\approx 5645^{+258129}_{-4811}$	DN/exp/pixel

Table C1. A summary of the unit conversions and sources of noise in the signal. Beyond the Photon Signal step, the values are interpreted as integers by the camera and readout systems.

Since there can be inherent variability in the count rate, we add Poisson noise. Likewise, the CCD converts the incoming photons to electrons based on the electron hole-pair Energy in Silicon ($E_{eh} = 5.847944 \cdot 10^{-12}$ erg/e$^-$). The CCD reads out the total electron count from each pixel in terms of voltage, with readout noise, gain, and gain bias. A separate calibration code interprets the voltage signal to the computer system as a Data Number based on this gain and gain-bias value. It is not necessary to include a gain-bias value for this testing, since this value doesn't fluctuate and is only present to prevent negative values in the computer processing system.

Appendix D. Inversion of Density Model

If we have enough confidence in the absorption cross-section, we can in practice recalculate the density model instead. Starting as before, we can take the ratio of the apogee image with any lower-altitude image as a proxy of the transmission factor. We then use a forward model to invert the problem and produce a potential solution for the atmospheric density profile. To do this, we set up an optimization routine to vary the parameters of an assumed functional form of the density profile.

This profile is carried through the spherical-shells model and compared with the simulated "measured" optical depth. The profile used here was formulated by observation of the NRL model to be a combination of a linear and parabolic line. Thus, using a conventional approach often used in setting up a broken power law (insert citation), we define the functional form of the density to roughly follow

$$y = \frac{f_1 - f_2}{\pi} \arctan\left(k\left(x_0 - x\right)\right) + \frac{f_1 + f_2}{2}, \tag{D.1}$$

where y is the density, x is the altitude, and $f_1 = a(x - x_0)^2 + b(x - x_0) + c$ and $f_2 = b(x - x_0) + c$ are the standard parabolic and linear functions, respectively. The coefficients a, b, c, and x_0 are adjusted by the optimization routine. The value for k adjusts the level of sharpness around the break-point between the two functions and is simply set to a relatively high number such as 1000.

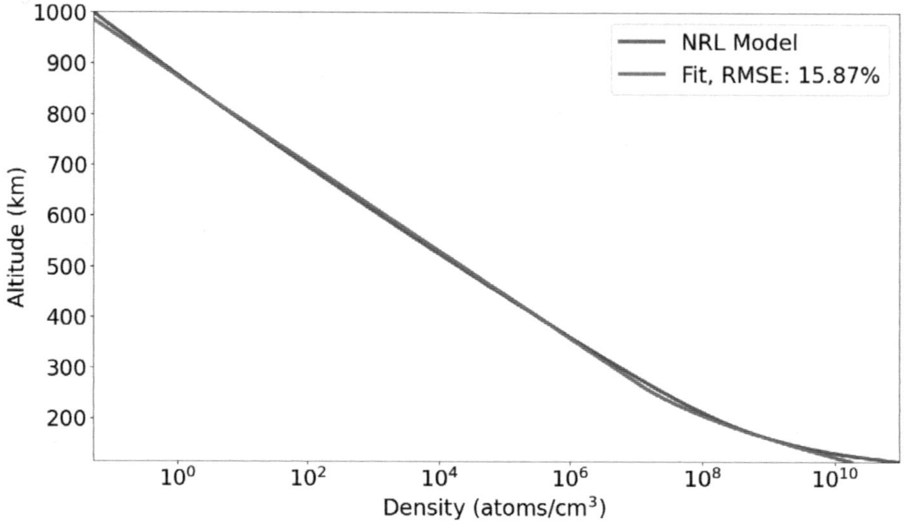

Figure D1. Comparing the density model as used to generate the simulated signal (green line) and the recomputed values through the optimization routine (red line).

An initial guess for the parameters is given based on a fitting of the NRL model. We then apply those parameters as the initial guess for computing the density within equation 1. This result is compared with the signal data, and the optimization routing minimizes the difference between the two by adjusting those initial guesses. As can be seen in Figure D1, the fit is not perfect but it is close. From these simulated noisy images we can reproduce the density curve from the NRL model between the observed altitude and 1000km with a Root Mean Squared Error (RMSE) of about 16%.

We do not consider this routine robust enough, as the initial guess is based on required prior knowledge and complete trust in the sparse absorption cross-section data set. This step will continue to be improved upon though, and combining observations from various shutters and timings may improve this routine. Additionally, we may be able to use the solved absorption cross-section from one image to inform this inversion of the data from another image.

Magnetic flux rope: What is it?

Qiang Hu
320 Sparkman Drive, CSPAR and Department of Space Science, The University of Alabama in Huntsville, Huntsville, AL 35805, USA

E-mail: qiang.hu@uah.edu

Abstract. Magnetic flux rope, a type of magnetic field structure in space plasmas, has been studied for decades through both observational and theoretical means. We provide a brief report on our recent modeling study of its magnetic field configuration based on in-situ spacecraft measurements, focusing on those made for large-scale flux ropes in the interplanetary space. We illustrate the complexity in its field-line topology by presenting two event studies employing a unique analysis method. In particular, we demonstrate the feasibility and challenges for the approach to use two or more in-situ spacecraft datasets. We discuss the implications of our results and offer some thoughts on further advancing the investigation of the nature of the magnetic flux rope.

1. Introduction

Magnetic flux rope is generally believed to form the core structure of a coronal mass ejection (CME) on the Sun upon its eruption. The evidence supporting this view becomes particularly strong when a CME is crossed in-situ by an observing spacecraft at larger heliocentric distances. Such crossings enable direct sampling of the plasma medium and the magnetic field embodied within a CME, thus the so-called interplanetary CME (ICME), yielding time-series measurements across a propagating ICME complex at a single point or along a single line. These observational signatures include but are not limited to, 1) enhanced magnetic field magnitude relative to the ambient solar wind, 2) smooth rotation in one or two field components, and 3) depressed proton β value (the ratio between the proton pressure and the magnetic pressure). A structure spanning an interval with all three aforementioned signatures present is classified as a magnetic cloud (MC) and is generally believed to possess a magnetic flux rope configuration [1, 2, 3]. Other ICME structures possessing two, one, or arguably none of those three signatures can still possess a magnetic flux rope configuration, especially when the correspondences to the source CME eruptions are certain. This generally calls for well coordinated observational analysis combining the modern remote-sensing observations of the Sun and adequate analysis methods taking into account the realistic complexity in the magnetic field topology of (I)CMEs (see, e.g., [4]).

Conceptually MCs are thought to possess well-organized magnetic field configuration since their first identification from in-situ spacecraft data. The most common model based on in-situ data assumes a cylindrical one-dimensional (1D) geometry [5, 1]. This type of configuration exhibits a field-line topology similar to a slinky and this general view has not evolved much over the past few decades. In this configuration, field lines are winding around a central axis on distinct surfaces with regularly shaped cross sections, i.e., mostly concentric circles or ellipses.

One significant extension to this type of models is a complete two-dimensional (2D) model based on the Grad-Shafranov (GS) equation, which yields a 2D arbitrary cross section directly derived from the spacecraft measurements along its path across the structure [6, 7, 8, 9]. One step further in the latest development is a quasi-three dimensional (3D) model that adds 3D spatial variations to a magnetic flux rope configuration [10, 11, 12]. It has been applied to a number of event studies, especially in connection with the solar source CME magnetic field topology that is characterized by more general 3D field-line configurations [13, 14, 4].

The question of what defines a magnetic flux rope remains, despite decades of observations and research. The exception, in our view, or one way to address this question is through multiple observations including remote-sensing imagery and numerical modeling of solar source regions of flares/CMEs. For example, a field-line twist number can be quantified and used to identify a flux rope structure by specifying a threshold value [15]. However it remains challenging for applying a unified approach to perform such numerical simulations in order to obtain reliable results [16]. For relevant modeling and analysis of in-situ measurements, it is equally challenging to achieve a high level of certainty. In this report, we do not intend to address this question in a definitive manner, but to demonstrate the necessity of going beyond the traditional 1D or 2D view of a magnetic flux rope configuration. Then one may achieve the goal of defining a magnetic flux rope in the most general as well as the simplest terms.

This report is organized as follows. Section 2 describes briefly the quasi-3D model for a magnetic flux rope and the optimal fitting approach. The set of analytic formulations is given. Section 3 presents two selected case studies by applying the quasi-3D model for the purpose of demonstrating the feasibility of the approach and the complexity in the magnetic flux rope configurations that may not be addressed by other models. Section 4 summarizes the case study results and discusses our view toward and the outlook for the analysis of magnetic flux ropes in space plasmas.

2. A quasi-3D model for magnetic clouds

It is highly desirable to move forward with a 3D model for the in-situ analysis of MCs to account for the reality of complex magnetic field topology clearly seen in countless studies of their solar source regions. One of such models has been developed recently by applying a set of formulations well known in fusion science [17] for a linear force-free magnetic field configuration. Namely, for the magnetic induction, $\mathbf{B}$, satisfying $\nabla \times \mathbf{B} = \mu\mathbf{B}$ with $\mu = Const$, the three components of the magnetic field can be written explicitly in a cylindrical coordinate system [17], (r, θ, z),

$$\frac{B_z(\mathbf{r})}{B_{z0}} = J_0(\mu r) + CJ_1(lr)\cos(\theta + kz), \tag{1}$$

$$\frac{B_\theta(\mathbf{r})}{B_{z0}} = J_1(\mu r) - \frac{C}{l}\left[\mu J_1'(lr) + \frac{k}{lr}J_1(lr)\right]\cos(\theta + kz), \tag{2}$$

$$\frac{B_r(\mathbf{r})}{B_{z0}} = -\frac{C}{l}\left[kJ_1'(lr) + \frac{\mu}{lr}J_1(lr)\right]\sin(\theta + kz). \tag{3}$$

Here the solution, dubbed Freidberg solution (FS), obviously has 3D spatial variations with the r dependence governed by the Bessel functions of the first kind. The constant parameters are C, k, and μ (with $l = \sqrt{\mu^2 - k^2}$), representing the amplitude of the non-axisymmetric part, a wavenumber along the z dimension, and the linear force-free constant, respectively. The magnetic field and spatial variables are normalized by a constant B_{z0} and a length scale a [10, 11], respectively.

A standard χ^2 minimization procedure following [18], as given below,

$$\chi^2 = \frac{1}{\mathtt{dof}} \sum_{\nu=X,Y,Z} \sum_{i=1}^{N} \frac{(b_{\nu i} - B_{\nu i})^2}{\sigma_i^2}, \tag{4}$$

Journal of Physics: Conference Series **2544** (2023) 012002 doi:10.1088/1742-6596/2544/1/012002

is devised to obtain a set of optimal parameters that determines a solution consistent with the spacecraft measurements along one or more paths across the solution domain [10, 11]. We name this approach the FS model fitting method. Here the reduced χ^2 value is evaluated along one single path or multiple separate spacecraft paths (for the latter, see [12]). Differences between the measured magnetic field **b** and the FS model output are evaluated component-wise, subject to the uncertainty estimates σ_i and the degree-of-freedom (`dof`), complying with the χ^2 statistics [18]. Detailed descriptions of the optimization procedures are given in [11].

One word of caution is that when performing this kind of optimization/minimization analysis, a proper evaluation of the uncertainties, σ_i, in the denominator of the formula inside the summation sign of equation (4) is critical for an objective assessment of the goodness-of-fit. If such uncertainties were arbitrarily enlarged (for example, by substituting the total magnetic field magnitude for this quantity), the principle of this analysis approach would be violated, rendering meaningless results.

Table 1. Locations of Solar Orbiter and Wind spacecraft on 30 May 2020, 00:00 UT in the Carrington coordinates (courtesy of Solar-MACH [19], `https://da.gd/gzoINr`).

Position	Solar Orbiter	Wind
R (AU)	0.56	1.0
longitude (°)	282	252
latitude (°)	4.1	-0.9

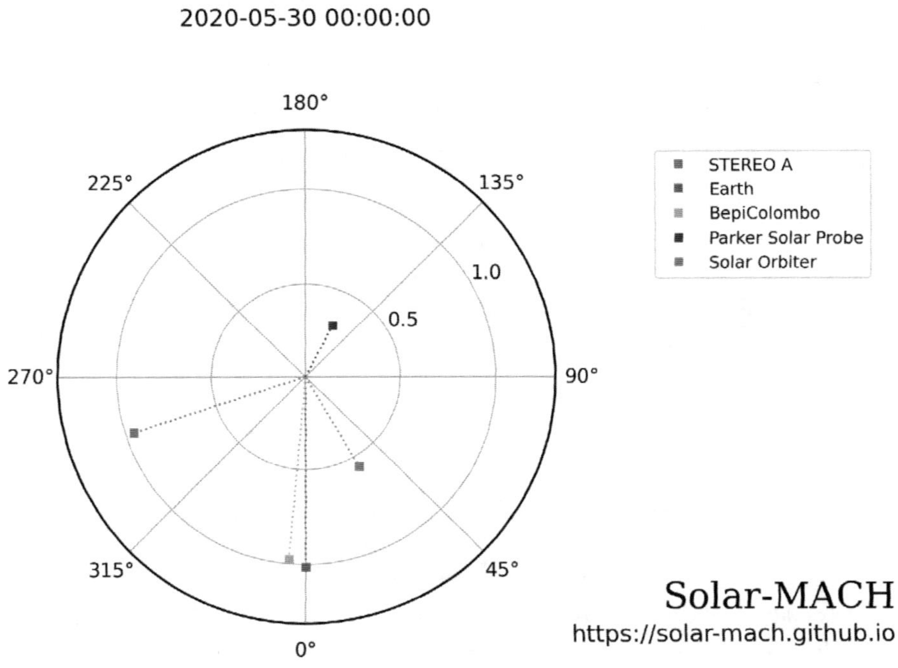

Figure 1. The Solar Orbiter and Wind spacecraft (Earth) locations on 30 May 2020 on the Sun's equatorial plane (see the legend and Table 1 for the corresponding coordinates).

3. Case studies: complexity in magnetic flux rope configurations

We present in this section two event studies to illustrate the complexity in the magnetic flux rope configurations by applying the FS model fitting approach to in-situ spacecraft data. By presenting these results, we hope to stimulate interest and the realization of the need to expand or reformulate the "conventional" perception about a magnetic flux rope.

3.1. Event 1: ICME on 30 May 2020

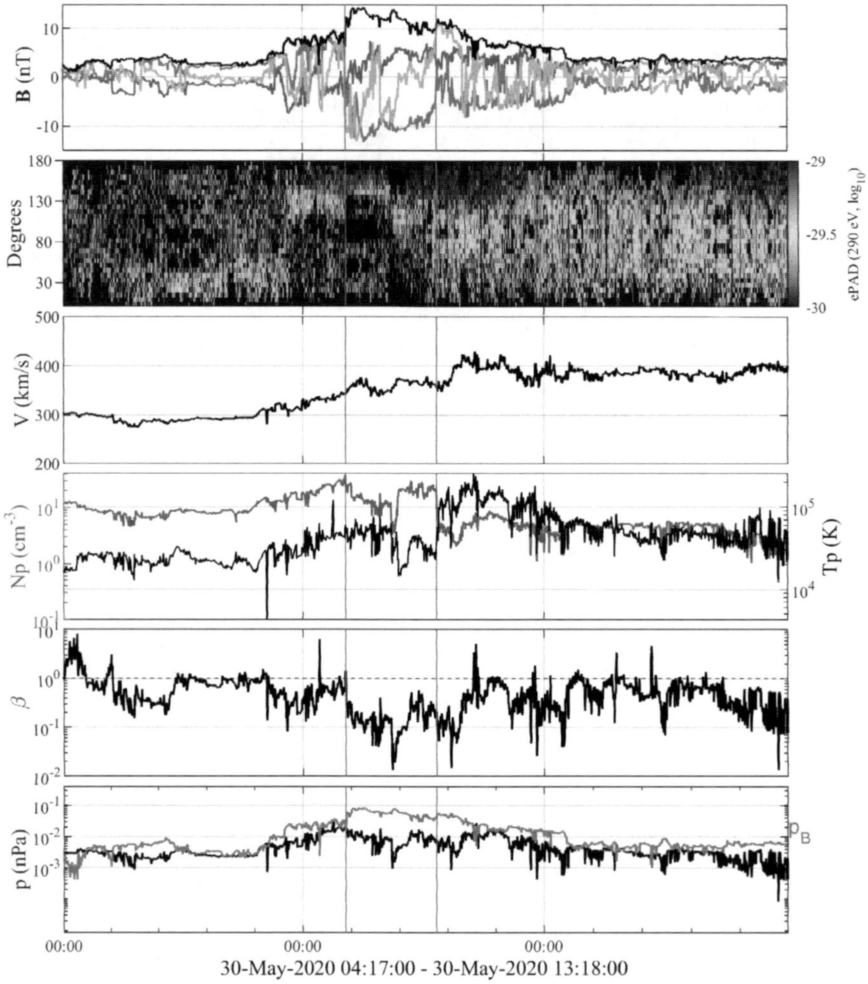

Figure 2. Time series from Wind spacecraft in-situ measurements for the time period, May 29 - June 1, 2020. From the top to bottom panels are: magnetic field components in GSE-X (blue), Y (red), and Z (gold) coordinates, and the magnitude (black), the electron pitch angle distributions (ePADs) with color scales given by the color bar to the right, the solar wind speed, the proton density and temperature, the proton β, and the proton and magnetic (red) pressure. The interval chosen for analysis is marked by the vertical red lines and is denoted beneath the last panel in UT.

Event 1 occurred during a time period with multi-spacecraft encounters in addition to the in-situ measurements by the Wind spacecraft at Earth. In particular the newly launched Solar

Orbiter (SO; [20]) was believed to have crossed the same ICME complex [21] at a heliocentric distance $R \approx 0.56$ AU with a longitudinal separation of 30° from Earth (see Table 1 for the spacecraft locations information). Figure 1 shows the locations of a number of spacecraft on the solar equatorial plane at the time. Three spacecraft, BepiColombo (BC), Wind and SO had intercepted the same ICME at different times due to their different locations as shown. The in-situ measurements from BC are not available to us and we can only obtain the SO magnetic field measurements for the time period we have examined. Figure 2 shows the usual set of time-series measurements from Wind spacecraft, including the ePADs data from the Solar Wind Experiment (SWE) instrument. The two vertical lines mark the interval selected for the FS model fitting method described in Section 2. The three signatures mentioned in Section 1 for an MC are present during this interval, supporting the following analysis by the FS model fitting. In addition, the ePADs show relatively prominent signatures of streaming electrons in the anti-field direction ($\sim 180°$ PA), which can also be confirmed by the similar data (not shown) by the 3D Plasma (3DP) instrument. Figure 3 shows the corresponding magnetic field measurements at SO obtained about 1 day earlier. At SO, the signatures of a magnetic flux rope structure are more prominent with stronger magnetic field magnitude and a more pronounced rotation in the B_N component during the interval between DOY 149.8 and DOY 150.4.

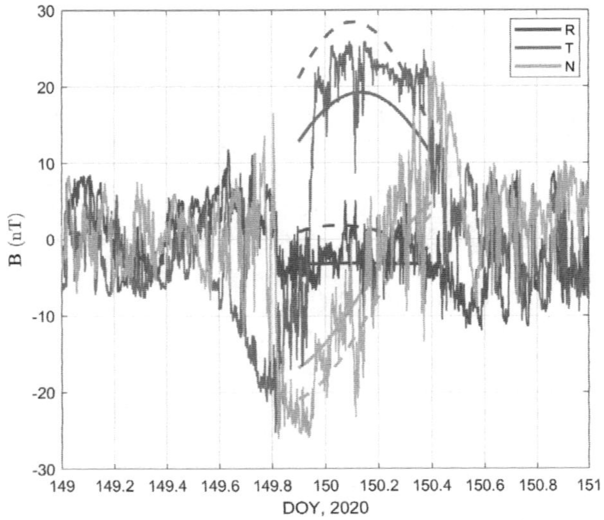

Figure 3. The SO magnetic field components in the spacecraft centered Radial (R), Tangential (T) and Normal (N) coordinates for the time period 28 - 30 May 2020 (day of year, DOY 149 - 151, 2020). The smooth curves are the FS model output discussed in Section 3.1.

We have demonstrated in recent studies [4, 12, 13] the necessity of employing more than one spacecraft dataset in the FS model fitting. This event study represents our latest effort and an extreme case in terms of the widest separation distance between the two spacecraft. Based on the assumption that both Wind and SO spacecraft encountered the same ICME structure (see also [21] for detailed discussions), we carry out the two-spacecraft dataset optimization approach [12] by employing both Wind and SO spacecraft data shown in Figures 2 and 3 for the FS model fitting. Figure 4 shows the optimal fitting result: left panel shows the fitting result of the FS model output and measured magnetic field components along the Wind spacecraft path across the analysis interval with the minimum reduced $\chi^2 = 1.86$; right panel shows one cross section of the FS model solution with the two spacecraft paths projected in this view down the z axis. It shows a configuration with one dominant positive polarity region enclosed by the boundary (dotted magenta line) where $B_z = 0$. The model output along the SO spacecraft path is denoted in Figure 3 by the smooth solid curves for the three components. The corresponding dashed curves are the estimates by assuming a radial evolution of the axial magnetic field B_z based on the consideration of the conservation of the axial magnetic flux content Φ_z [13].

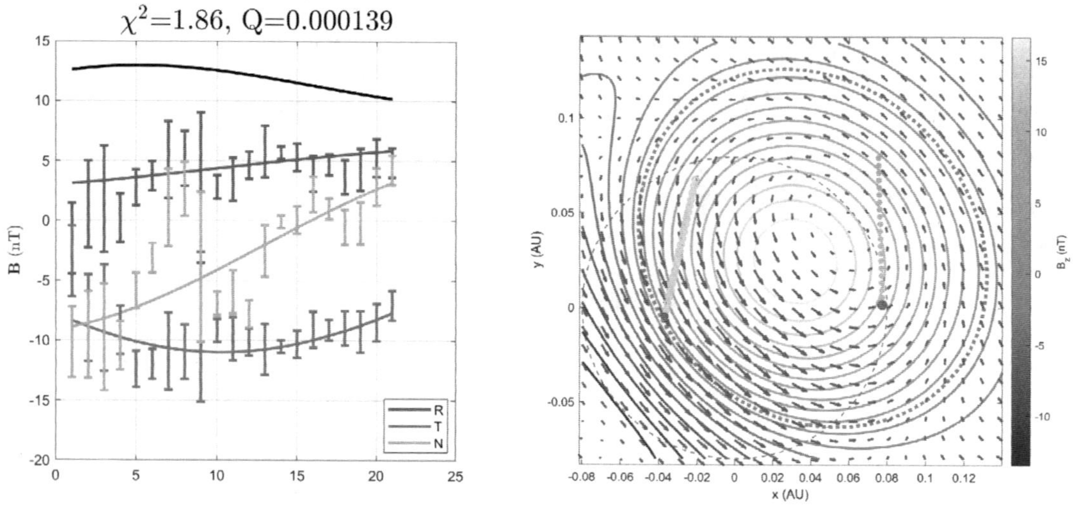

Figure 4. The FS model fitting result to the combined Wind and SO spacecraft datasets. Left panel: the optimal model fitted (smooth curves) and measured (errorbars) magnetic field components along the Wind spacecraft path; black curve is the fitted field magnitude. Right panel: the cross section of the magnetic field configuration at $z = 0$ from the fitted FS model; the contours are the axial field component B_z with scales indicated by the color bar, the arrows are the transverse field vectors, and the Wind and SO spacecraft paths are denoted by the points to the right, color coded by the B_z component, and the green points to the left, respectively; the dotted magenta line marks the boundary of the positive polarity region where $B_z = 0$.

Figure 5 puts the whole FS model solution domain in a view that better reveals the flux rope configuration from the encountering Wind spacecraft's perspective in space. The main flux bundle (orange lines) is winding mostly in the East-West direction and the SO spacecraft has a path across the middle while the Wind spacecraft path appears to cross the edge to the South. As a check for this type of analysis using two sets of in-situ spacecraft data, one can further examine the correlation between the FS model output and the actual measured magnetic field components along the secondary spacecraft's path, i.e., SO's path in this case. Such a comparison is shown in Figure 6. The overall correlation coefficient between the two sets of magnetic field components is $cc = 0.94$.

One caveat, however, is that a simple timing analysis based on an unknown but assumed constant average traveling speed between SO and Wind fails to yield a reasonable time delay consistent with the observations at the two spacecraft as shown above. If the whole configuration demonstrated in Figure 5 is believed to be reasonable based on the current analysis, then a highly variable kinematic evolution between the two spacecraft has to occur, which has yet to be verified based on current available observations. The processes unaccounted for between two or more observing probes pose significant challenges for this type of analysis, especially when

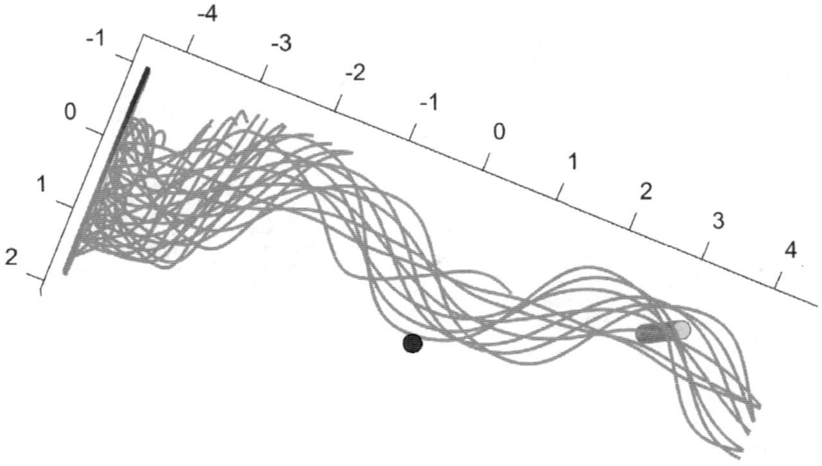

Figure 5. The 3D view of the FS model fitted field-line configuration along the Wind spacecraft path (blue dot) toward the Sun. The East-West direction is horizontal and North is upward. The orange lines are the field lines originating from the cross section shown to the East. The SO spacecraft path (green dots) across the same structure is shown to the West. The tickmark labels are normalized by a length scale $a = 0.080$ AU.

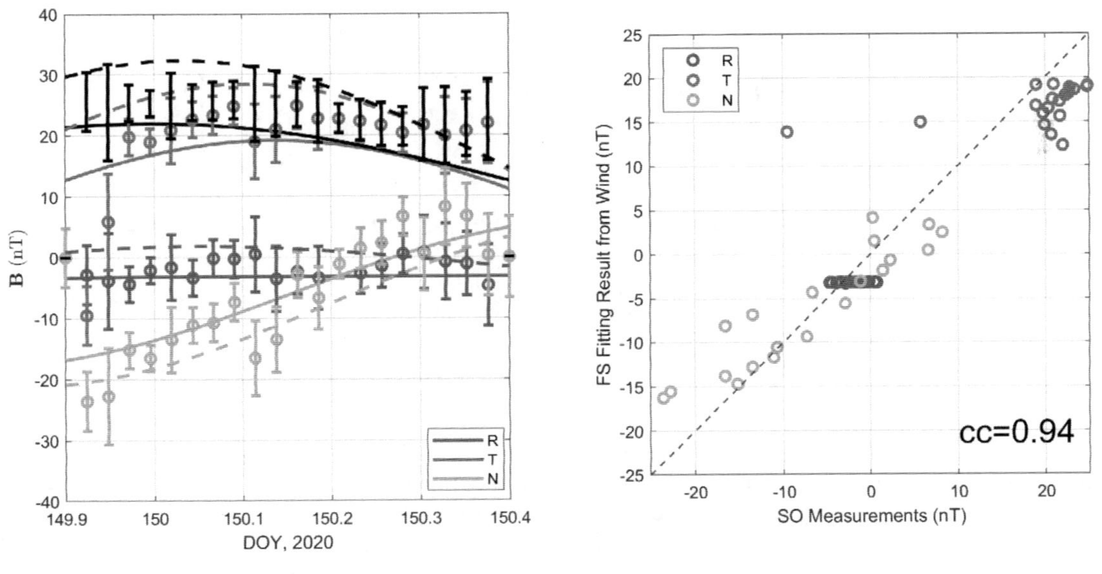

Figure 6. The correlation between the measured magnetic field components (circles with errorbars) and the FS model fitting results (smooth curves) along the SO spacecraft path based on the 3D configuration shown in Figures 4 and 5. See Section 3.1 for details.

the separations are relatively large as they are for this case.

3.2. Event 2: ICME on 9 March 2012

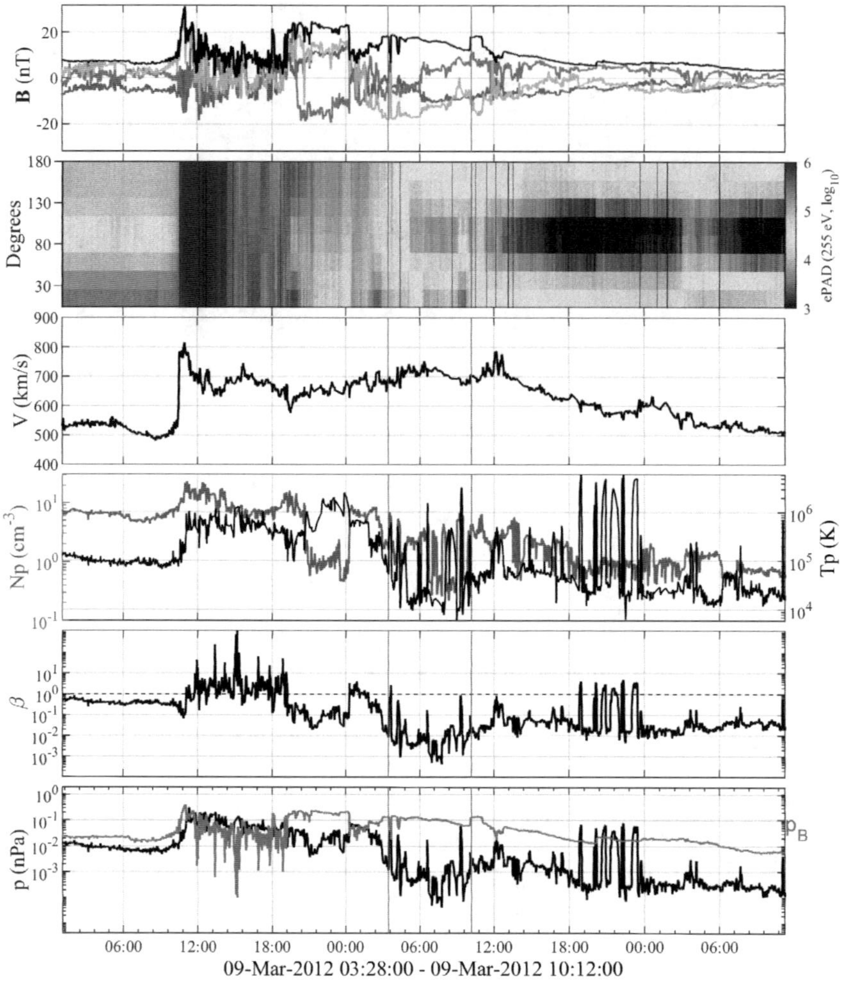

Figure 7. Time series from the Wind spacecraft in-situ measurements for the time period March 8 - 11, 2012. Format is the same as Figure 2. The second panel shows the ePADs from the Wind/3DP instrument.

To further illustrate the level of complexity as revealed from in-situ modeling of ICME magnetic flux ropes, we present one more analysis of the ICME event on 9 March 2012. Figure 7 shows again the time-series measurements from the Wind spacecraft. An ICME complex is seen preceded by a shock wave with a turbulent sheath region. The analysis interval with relatively strong and smooth magnetic field falls behind, characterized by significantly reduced proton β value (in the order of 0.01) and relatively steady solar wind speed. The ePADs data from Wind/3DP (the corresponding SWE data have significant gaps) show a slight hint of bi-directional electrons streaming mainly along the 0° and 180° PA directions during the interval. Figure 8 shows the usual output for the optimal FS model fitting by using the Wind spacecraft data from the selected interval. The result is acceptable as indicated by the minimum reduced $\chi^2 = 1.26$ and the corresponding goodness-of-fit parameter Q [18, 11]. The cross section perpendicular to $\hat{\mathbf{z}}$ shows two major polarities with opposite signs of the B_z component. The

Journal of Physics: Conference Series **2544** (2023) 012002 doi:10.1088/1742-6596/2544/1/012002

spacecraft path appears to be embedded entirely inside the flux bundle of the negative polarity. This is confirmed by a rendering of the 3D field-line configuration in Figure 9 with a viewpoint from the ecliptic north onto the ecliptic plane. The overall field-line configuration exhibits two flux bundles winding around each other that are spiraling mainly along the Sun-Earth direction and nearly lying on the ecliptic plane, but are oppositely directed. The Wind spacecraft path crosses the flux bundle in cyan color in the left panel of Figure 9, where the cyan field lines are all pointing downward (anti-sunward) in this view. The orange color field lines, on the other hand, are all pointing upward. In the right panel of Figure 9, only selected field lines intercepting the Wind spacecraft path and a pair of field lines originating near the location of the maximum B_z in the positive polarity region are highlighted in blue and red colors, respectively. A number of blue field lines disperses widely with the footpoints spreading irregularly on the bottom cross section as shown, although the overall appearance may still resemble a "tight" spiral shape as commonly perceived.

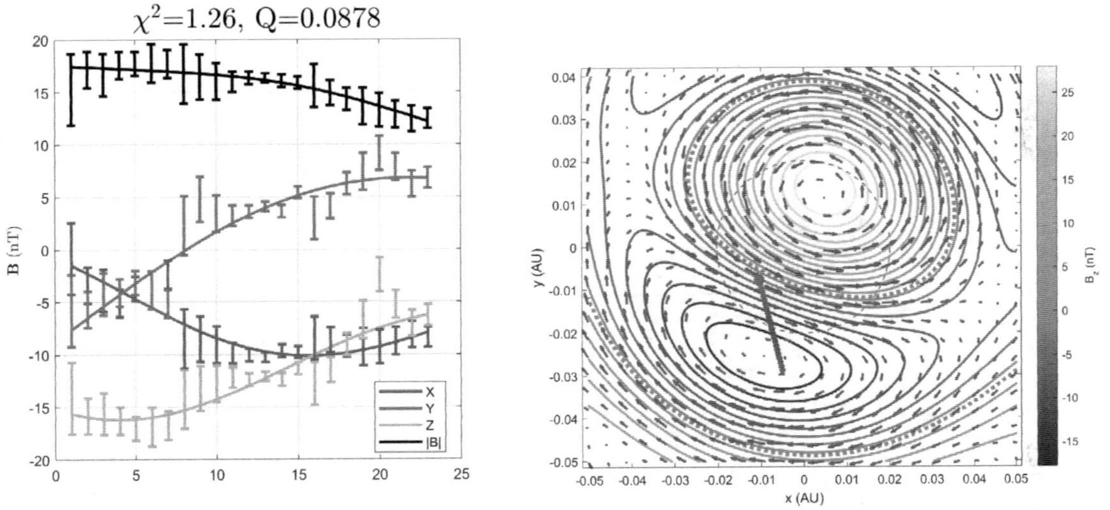

Figure 8. The optimal fitting result for Event 2 on 9 March 2012 by using the Wind spacecraft in-situ measurements. Format is the same as Figure 4.

One more aspect as demonstrated in Figure 9 is that the Wind spacecraft path misses the main body of the flux bundle illustrated by the orange field lines rooted in the positive polarity region on the cross-section plane. Whether this flux bundle is part of one entire ICME flux system has yet to be determined by examining the solar source region eruption(s) of the corresponding flare(s) and/or CME(s).

4. Summary and discussion: "food for thought"

In summary, we have presented two case studies of ICME magnetic flux ropes by using the quasi-3D FS model fitting approach to the in-situ spacecraft measurements. Their main parameters are summarized in Table 2. They show a typical range of magnitudes of the axial magnetic flux contents Φ_z (in the order of tera Wb, TWb) similar to prior results obtained by using the 2D GS reconstruction method [22]. For Event 2, the axial magnetic flux contents for both the positive (+) and negative (-) polarity regions are given and they are close in magnitudes. However the 3D spatial features such as the clear variation along the axial (z) dimension are distinct from other 1D or 2D models (see also [4, 14, 10]). Therefore one needs to keep an open mind as

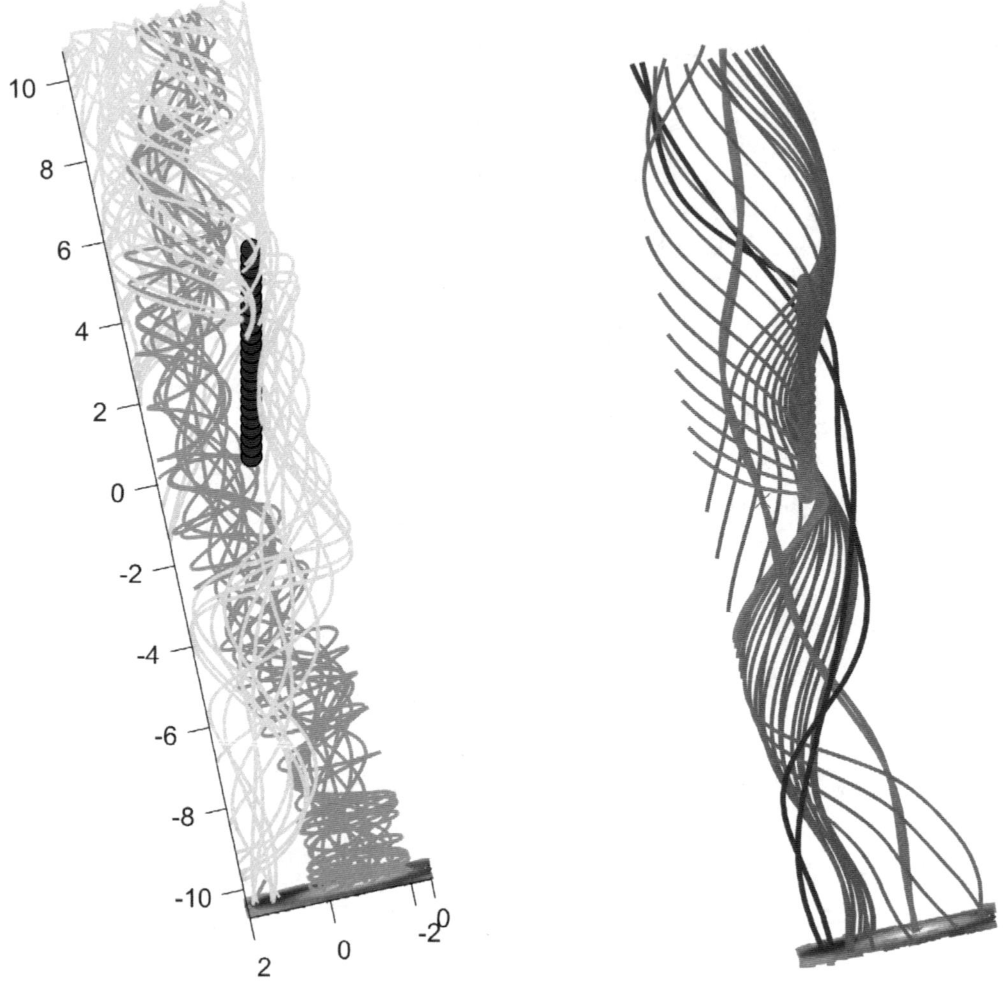

Figure 9. Left panel: the 3D view of the field-line configuration of Event 2 down from the ecliptic North (Sun is in the upward direction). The orange and cyan lines are field lines rooted on the positive and negative polarity regions on the cross-section plane in the bottom. The blue dots form the Wind spacecraft path for this analysis interval. Tickmark labels are normalized by a length scale $a = 0.020$ AU. Right panel: the same view as the left panel but with only selected field lines intercepting the Wind spacecraft path color coded by the B_z component (all negative), while the red lines are rooted near the maximum B_z of the positive polarity region.

far as a definition of a "magnetic flux rope" is concerned. Our current view is that from the perspective of terminology, the historic term "magnetic clouds" may be used interchangeably with "magnetic flux ropes". A universal definition of a magnetic flux rope can be sought from solar observations and analysis, especially concerning the CME magnetic field topology. Then the question naturally becomes whether a magnetic flux rope always constitutes an essential part of the structure of a CME. The answer may be positive if one adopts a broader definition of a magnetic flux rope, e.g., that of a "ribbon-like" field-line topology [4, 12].

Table 2. Summary of primary FS model parameters for the two events.

Parameters	$\hat{\mathbf{z}}$ (GSE)	CB_{z0} (nT)	ka	μa	Φ_z (TWb)
Event 1	[0.022, -0.92, -0.38]	22	1.4	2.4	+4.5
Event 2	[0.99, 0.11, 0.086]	-38	-0.29	1.8	-1.0/+0.83

One may add to the title of this report the additional question, "What is it good for?". As widely recognized, the importance and prevalence of a magnetic flux rope structure to CMEs are well established. It is generally believed that flux rope formation prior to and during flare/CME eruptions plays a critical role in the associated dynamic evolution processes. The CME flux rope configurations have been well manifested in in-situ observations of ICMEs, which occupy the far-right end portion of the scale-size distribution of magnetic flux ropes more extensively identified over a wide range of scale sizes [23]. The so-called small to intermediate-scale magnetic flux ropes are found to be much more prevalent in the interplanetary space and their scale sizes exhibit a continuous power-law distribution [23]. They are intrinsically related to the magnetohydrodynamic turbulence and play active roles in particle energization and heating of plasmas, based on recent studies [24, 25], a few also included in this volume. For those topics beyond the scope of this report, we refer interested readers to those works, and in particular, to an online database of small-scale magnetic flux ropes at `http://fluxrope.info`, for event lists from multiple spacecraft missions and the relevant references. Last but not the least, the Earth's magnetospheric environment or more generally the planetary environments provide another space plasma regime for exploring the nature of magnetic flux ropes.

Acknowledgments
We thank Dr. Gary Zank (Co-I of the SO mission) and Dr. Lingling Zhao for providing the SO magnetic field data for this study. The author also benefited from the on-going collaboration with Prof. Jiong Qiu at the Montana State University. We acknowledge support from NASA grants 80NSSC19K0276, 80NSSC21K0003, 80NSSC21K1671, and NSF grants AGS-1954503, AGS-2020703, and AGS-2050340. The Wind spacecraft data are downloaded from the NASA CDAWeb and OMNIWeb Plus.

References

[1] Burlaga L F, Klein L, Sheeley N R J, Michels D J, Howard R A, Koomen M J, Schwenn R and Rosenbauer H 1982 *Geophys. Res. Lett.* **9** 1317–1320
[2] Burlaga L F 1988 *Journal of Geophysical Research: Space Physics* **93** 7217–7224 URL `https://agupubs.onlinelibrary.wiley.com/doi/abs/10.1029/JA093iA07p07217`
[3] Lepping R P, Burlaga L F and Jones J A 1990 *J. Geophys. Res.* **95** 11957–11965
[4] Hu Q, Zhu C, He W, Qiu J, Jian L K and Prasad A 2022 *Astrophys. J.* **934** 50 (*Preprint* `2204.03457`)
[5] Lundquist S 1950 *Ark. Fys.* **2** 361
[6] Hu Q 2017 *Sci. China Earth Sciences* **60** 1466–1494
[7] Hau L N and Sonnerup B U Ö 1999 *J. Geophys. Res.* **104** 6899–6918
[8] Hu Q and Sonnerup B U Ö 2001 *Geophys. Res. Lett.* **28** 467–470
[9] Hu Q and Sonnerup B U Ö 2002 *J. Geophys. Res. (Space Physics)* **107** 1142
[10] Hu Q, He W, Qiu J, Vourlidas A and Zhu C 2021 *Geo. R. Letts.* **48** e90630 (*Preprint* `2010.11889`)
[11] Hu Q 2021 *Solar Phys.* **296** 101 (*Preprint* `2104.09352`)
[12] Hu Q, He W and Chen Y 2022 *Frontiers in Physics* **10** 960315 (*Preprint* `2206.01247`)
[13] Hu Q, He W, Zhao L and Lu E 2021 *Frontiers in Physics* **9** 407 ISSN 2296-424X URL `https://www.frontiersin.org/article/10.3389/fphy.2021.706056`

[14] He W, Hu Q, Jiang C, Qiu J and Prasad A 2022 *Astrophys. J.* **934** 103 (*Preprint* 2201.03149)
[15] Liu R, Kliem B, Titov V S, Chen J, Wang Y, Wang H, Liu C, Xu Y and Wiegelmann T 2016 *The Astrophysical Journal* **818** 148 URL https://doi.org/10.3847/0004-637x/818/2/148
[16] Jiang C, Feng X, Guo Y and Hu Q 2022 *The Innovation* **3** 100236 ISSN 2666-6758 URL https://www.sciencedirect.com/science/article/pii/S2666675822000327
[17] Freidberg J P 2014 *Ideal MHD* (Cambridge, UK: Cambridge University Press) pp 546–547
[18] Press W H, Teukolsky S A, Vetterling W T and Flannery B P 2007 *Numerical Recipes in C++ : The Art of Scientific Computing* (New York: 778, Cambridge Univ. Press)
[19] Gieseler J, Dresing N, Palmroos C, von Forstner J L F, Price D J, Vainio R, Kouloumvakos A, Rodríguez-García L, Trotta D, Génot V, Masson A, Roth M and Veronig A 2022 Solar-mach: An open-source tool to analyze solar magnetic connection configurations URL https://arxiv.org/abs/2210.00819
[20] Müller, D, St Cyr, O C, Zouganelis, I, Gilbert, H R, Marsden, R, Nieves-Chinchilla, T, Antonucci, E, Auchère, F, Berghmans, D, Horbury, T S, Howard, R A, Krucker, S, Maksimovic, M, Owen, C J, Rochus, P, Rodriguez-Pacheco, J, Romoli, M, Solanki, S K, Bruno, R, Carlsson, M, Fludra, A, Harra, L, Hassler, D M, Livi, S, Louarn, P, Peter, H, Schühle, U, Teriaca, L, del Toro Iniesta, J C, Wimmer-Schweingruber, R F, Marsch, E, Velli, M, De Groof, A, Walsh, A and Williams, D 2020 *A&A* **642** A1 URL https://doi.org/10.1051/0004-6361/202038467
[21] Weiss A J, Möstl C, Davies E E, Amerstorfer T, Bauer M, Hinterreiter J, Reiss M A, Bailey R L, Horbury T S, O'Brien H, Evans V, Angelini V, Heyner D, Richter I, Auster H U, Magnes W, Fischer D and Baumjohann W 2021 *Astron. Astrophys.* **656** A13 (*Preprint* 2103.16187)
[22] Hu Q, Qiu J, Dasgupta B, Khare A and Webb G M 2014 *Astrophys. J.* **793** 53 (*Preprint* 1408.1470)
[23] Hu Q, Zheng J, Chen Y, le Roux J and Zhao L 2018 *The Astrophysical Journal Supplement Series* **239** 12 URL https://dx.doi.org/10.3847/1538-4365/aae57d
[24] le Roux J A and Zank G P 2021 *Astrophys. J.* **913** 84
[25] Zank G P, Nakanotani M, Zhao L L, Du S, Adhikari L, Che H and le Roux J A 2021 *Astrophys. J.* **913** 127

Magnetic Topology and the Structure of the Slow Solar Wind

M Asgari-Targhi[1], A Asgari-Targhi[1,2] and S R Habbal[3]

[1]High Energy Division, Harvard-Smithsonian Center for Astrophysics, 60 Garden Street, Cambridge, MA 02138, USA
[2]Harvard Faculty of Arts and Sciences
[3]Institute for Astronomy, University of Hawaii, 2680 Woodlawn Drive, Honolulu, HI, USA

E-mail: masgari-targhi@cfa.harvard.edu, aasgari@fas.harvard.edu, habbal@hawaii.edu

Abstract.
We present a 3D reduced magnetohydrodynamic (RMHD) model of reflection driven Alfvén wave turbulence in an open magnetic field positioned near the solar equator. The non-linear interactions between outward and inward propagating waves generate turbulence. The RMHD equations describing the turbulence include the effects of solar wind outflow velocity on the dissipation of waves. For Alfvén wave turbulence to be a viable mechanism for heating the corona and accelerating the slow solar wind, there must be sufficient counter-propagating waves to generate the turbulence. In our previous study of the fast wind, we showed that the Alfvén wave turbulence model creates the energy needed for accelerating the fast solar wind, when observed density fluctuations are included in the model. Here, we will explore whether the conditions of the equatorial corona, a source of slow solar wind, are capable of generating sufficient Alfvén wave turbulence to account for the acceleration of the slow wind. We show that for a specific sets of model parameters, the energy from the Alfvén wave turbulence model is smaller by an order of magnitude than the energy needed to heat and accelerate the slow solar wind.

1. Introduction

Alfvén wave turbulence is theorized to be a key process for depositing energy in the corona and accelerating the solar wind. The solar wind is broadly categorized as fast or slow. The fast solar wind has velocities of 500–800 $\mathrm{km\,s^{-1}}$ [1] and originates from coronal holes [2–8]. For the fast wind, observational and modeling studies support wave-turbulence-driven models [9–14]. Observed signatures of the wave turbulence continue out to 1 AU, where in situ observations find that the fast wind has a high Alfvénicity, that is, there is a strong correlation between the turbulent magnetic and velocity fluctuations. However, the origins and acceleration for the slow solar wind remain unclear.

The properties of the slow solar wind are highly variable, and velocity alone is insufficient to fully categorize the slow wind. However, typically slow solar wind has velocities ≤ 500 $\mathrm{km\,s^{-1}}$. The slow wind is often found near the solar equator during solar minimum and is more widely dispersed in latitude during more active periods. Slow wind usually has lower proton temperatures and higher densities than the fast wind [15]. Charge state abundances are consistent with sources that have a higher temperature than fast wind sources and the elemental composition of the slow wind suggests that the material was confined to closed loops before being released [16, 17]. Based on these factors, some slow solar wind likely originates from the boundaries of equatorial coronal holes or the low latitude extensions of polar coronal holes [e.g., 18].

The fundamental dissipation mechanism for the Alfvén waves is considered to be wave turbulence, driven by the interaction of inward and outward propagating waves via a well-known nonlinear process [19–21]. These nonlinear interactions produce turbulence and result in a rapid transfer of wave energy to smaller and smaller spatial scales in the direction transverse to the background field. The wave energy is then converted into the heat of the background plasma and is dissipated. In this paper, we carry out an observationally constrained modeling study to understand how Alfvén wave turbulence heats the corona and accelerates the slow solar wind.

2. Observations

2.1. Magnetic Field

The active region used in the present study is NOAA 10978, which crossed the central meridian on 2007 December 12 and has a bipolar magnetic structure. We used the Stereo A [22] data taken around 06:35 UT on that day. Figure 1 shows the magnetic flux distribution in the photosphere (red and green contours) superimposed on the 195 Å image from the heliospheric imager (HI) [23] on board Stereo A. The flux distribution is derived from the line-of-sight magnetogram taken on 2007 December 09 at 06:23:30 UT by the Michelson Doppler Imager (MDI) on board the Solar and Heliospheric Observatory [SOHO; 24]. There are sunspots present in this region, and the HI image shows coronal loops, fans, and open fields emanating from the two spots and the surrounding areas.

This active region is positioned near the equator and does not exhibit any twist or sigmoidal structures, so it is modeled using the Potential Field Source Surface (PFSS) model. We construct a PFSS model of this region using the flux distribution deduced from the MDI magnetogram and the SOLIS synoptic map for Carrington rotation 2064. The PFSS was computed using the Coronal Modeling System (CMS) software [25–28]. We trace 18 randomly selected field lines through the PFSS model. Figure 1(a) shows the selected field lines superposed on the HI 195 Å image.

The CMS constructs a realistic model of the coronal magnetic field at low heights $< 3R_\odot$ [27]. However, for the wave turbulence model, we need to include the magnetic field measurements to heights of 20 solar radii. Therefore, we apply an analytical model and combine the results with the CMS program to measure the magnetic field from 1-20 $R_\odot$.

In the analytical model, we solve $\nabla^2\Psi = 0$ for Ψ using a Legendre polynomial expansion. Thus,

$$\Psi(r,\theta) = R\sum_{m=0}^{\infty}\left(\frac{r}{R}\right)^{-(m+1)} a_m P_m(\cos\theta); \tag{1}$$

$$B_r(r,\theta) = \sum_{m=0}^{\infty}(m+1)\left(\frac{r}{R}\right)^{-(m+2)} a_m P_m(\cos\theta). \tag{2}$$

Given $B_r = \cos^8\theta$ only even harmonics survive. Letting $m = 2(n-1)$ and using the orthogonality of the Legendre polynomials, we find

$$B_r(r,\theta) = \sum_{n=1}^{5} B_n\left(\frac{r}{R_\odot}\right)^{-2n} P_{2(n-1)}(\cos\theta), \tag{3}$$

where $B_n = B_{\rm pole}[715, 2600, 2160, 832, 128]/6435$.

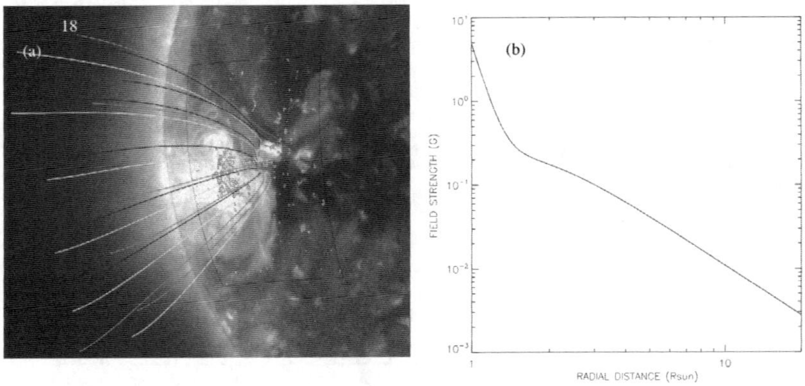

Figure 1. (a) An active region close to the equator and to the East limb observed by Stereo A on 2007 December 09 at 06:05:30 UT. Open fields have been randomly selected using the potential field extrapolation. Different colors are chosen to aid visualization. The positive and negative polarities on the solar surface are shown in red and green. (b) The magnetic variation as a function of position along the open field line, numbered 18 from (a).

For field lines in the equatorial plane, where the slow solar wind originates from, we have $\theta = \pi/2$, and the coefficients are $B_n = B_{\rm pole}[715, -1300, 810, -260, 35]/6435$ with $n = 1, \cdots, 5$. The field strength $B_0(r)$ inside the flux tube is assumed to be equal to that of the background field.

For one of the selected field lines, numbered 18 in Figure 1(a), we determine the magnetic field strength $B(s)$ as a function of position s along the field line. We combine the magnetic field measurements from the CMS program and the analytical model. The resulting curve is shown in Figure 1(b). The magnetic field is needed as an input for the wave modeling described in the next section.

2.2. Density and Velocity

[15] studied the solar wind speed associated with the Fe average charge states in the 2006 events during the descending Solar Cycle 23 using observations of SWEPAM-SWICS on ACE. Figure 2 shows the 2-hour time-averaged Fe charge state, and the corresponding solar wind speed was taken for 12 months in 2006. In our RMHD model of the slow wind presented in the next section, we incorporated the increase in density for the slow wind compared to the fast wind. The outflow velocity in our model is consistent with the slow wind velocity ≤ 500 km s^{-1} shown in Figure 2.

3. Reduced MHD Turbulence

Our model considers a flux tube that is positioned near the equator and extends from the coronal base outward into the heliosphere. The base is assumed to be located at a radial distance of $r_{\rm base} \approx 1\ R_\odot$. We follow the open field out to $r_{\rm max} = 20\ R_\odot$. At the base of the open field, the footpoint motions with velocity amplitude $v_{\rm rms} \approx 40$ kms^{-1} generate Alfvén waves. The observations of non-thermal velocity are consistent with the footpoint velocity incorporated in our wave turbulence model.

We use the RMHD equations [29, 30] to simulate the dynamics of Alfvén waves in a slow wind. The RMHD approximation studies the dynamics of the waves in a plane perpendicular to the open field axis and assumes that the magnetic field fluctuations are small compared to the background field. RMHD accurately describes Alfvén-waves turbulence and has been widely used in the study of turbulence [5, 12, 13, 26, 31–33].

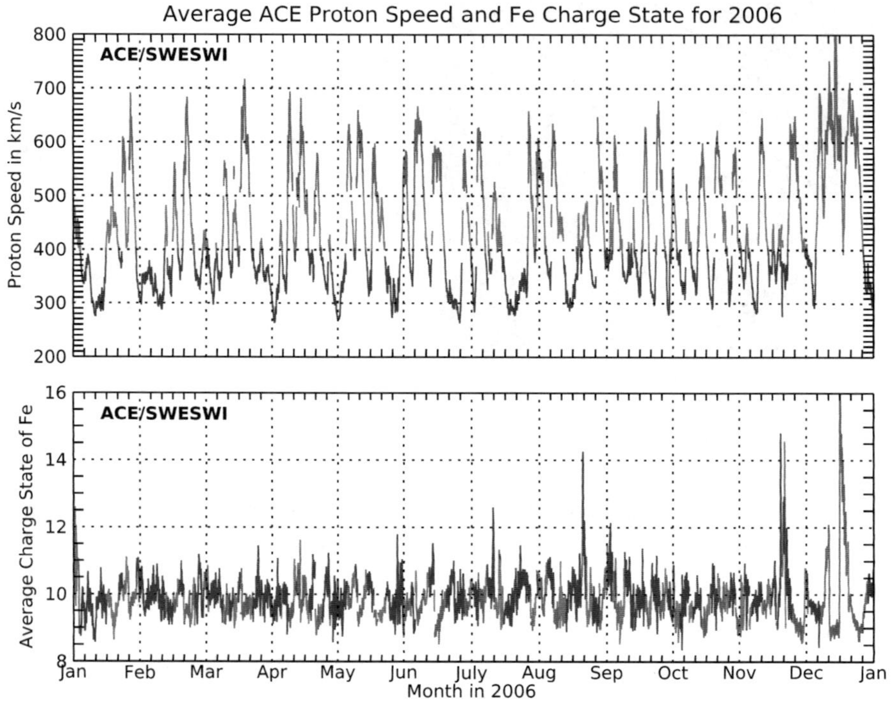

Figure 2. The 2 hr time-averaged Fe charge state and corresponding solar wind speed taken from the SWEPAM-SWICS data on ACE for 12 months in 2006 during the declining phase of Sunspot Cycle 23. Blue represents data for speeds below 400 kms^{-1}, red represents speeds in the range of 400?500 kms^{-1}, and green represents speeds greater than 500 kms^{-1} .

In the RMHD, the non-linear interactions between the inward Z_- and outward Z_+ propagating waves generate turbulence. The waves are described by Elsasser variables

$$\mathbf{z}_\pm \equiv \mathbf{v}_1 \mp \mathbf{B}_1/\sqrt{4\pi\rho_0} \tag{4}$$

where $\mathbf{B}_1$ and $\mathbf{v}_1$ are the magnetic and velocity fluctuations of the waves, and ρ_0 is the mean plasma density. In this picture, although the non-linear processes among the Alfven modes are considered, other non-linear effects, such as the coupling with fast- and slow mode waves are ignored. The effects of the slow wind outflow velocity, the magnetic field expansion, and the density fluctuations are included in the RMHD equations and are described by

$$\begin{aligned}\frac{\partial\omega_\pm}{\partial t} &= -(u_0 \pm v_\mathrm{A})\frac{\partial\omega_\pm}{\partial r} \\ &+\frac{1}{2}\left(\frac{dv_\mathrm{A}}{dr} \pm \frac{u_0}{2H_\rho}\right)(\omega_+ - \omega_-) + \frac{u_0}{2H_\mathrm{B}}(\omega_+ + \omega_-) \\ &-[\omega_+, f_-] - [\omega_-, f_+] \pm \nabla_\perp^2\left([f_+, f_-]\right),\end{aligned} \tag{5}$$

where $\omega_\pm \equiv -\nabla_\perp^2 f_\pm$ are the vorticities associated with the outward and inward waves, $u_0(r)$ is the outflow velocity of the wind, $v_\mathrm{A}(r)$ is the Alfvén speed, $H_B(r) \equiv B_0/(dB_0/dr)$ is the magnetic scale length, $H_\rho(r) \equiv \rho_0/(d\rho_0/dr)$ is the density scale length, and $B_0(r)$ is the magnetic field strength. The bracket term comprises of the non-linear terms and is defined by

$$[a, b] \equiv \frac{\partial a}{\partial x}\frac{\partial b}{\partial y} - \frac{\partial a}{\partial y}\frac{\partial b}{\partial x}, \tag{6}$$

where x and y are the coordinates perpendicular to the flux tube axis. The detailed derivation of the RMHD equations for a fast solar wind originating from a thin flux tube with a circular cross section, variable field strength $B_0(r)$ and density $\rho_0(r)$ are given in [12]. The RMHD equations describing both fast and slow solar wind are similar. The distinguishing factors between the two environments are the magnetic structure, density, and outflow velocity.

3.1. Modeling Results

Three-dimensional (3D) RMHD modeling requires that we first set up a one-dimensional (1D) background atmosphere inside the open field. One input parameter of our model is the open magnetic field, which is determined using the method described in the previous section.

The tube is assumed to have a circular cross-section with radius $R(r)$. Magnetic flux is conserved, so $R^2 B_0$ is constant along the tube. The radius of the flux tube at the coronal base is assumed to be $R_{\rm base} = 1$ Mm, which implies that the tube is everywhere thin.

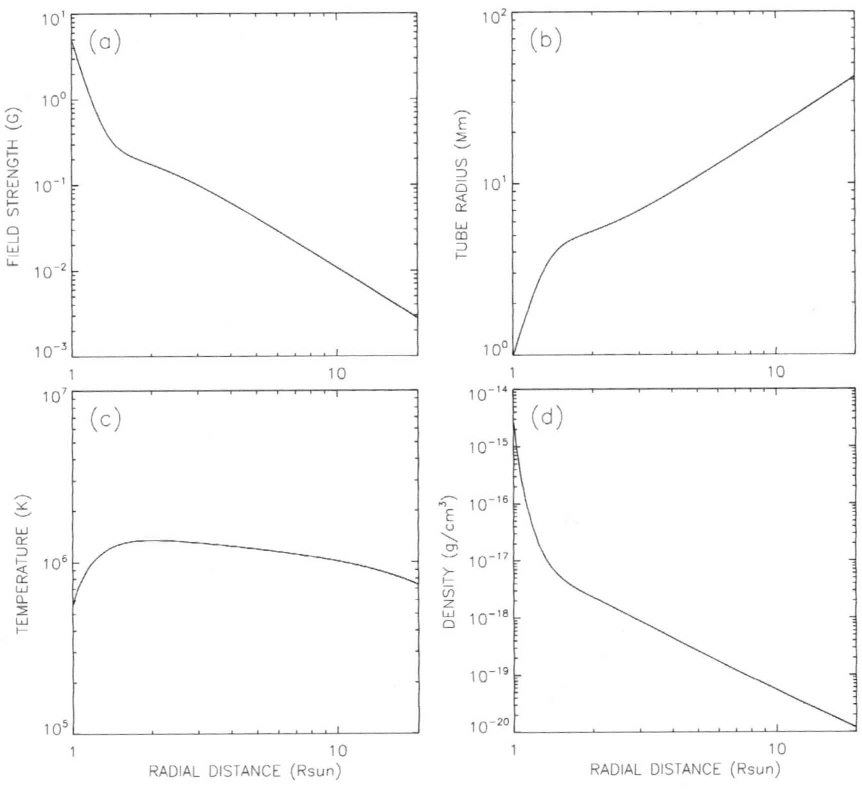

Figure 3. Radial dependence of various background quantities for an equatorial open field: (a) magnetic field strength $B_0(r)$, (b) flux tube radius $R(r)$, (c) temperature $T_0(r)$, and (d) density $\rho_0(r)$.

Figure 3 shows the background atmosphere structure. Figure 3(a) shows the magnetic field strength $B_0(r)$. Figure 3(b) shows the flux tube radius $R(r)$, which is computed from flux conservation. Note that $R(r)$ increases from 1 Mm at the coronal base to 60 Mm at $r = 20\ R_\odot$. Figure 3(c) shows the temperature $T_0(r)$, which increases from ~ 0.3 MK at the coronal base, and reaches a maximum of ~ 1.31 MK at $r \approx 3\ R_\odot$. Figure 3(d) shows the corresponding density $\rho_0(r)$, which is calculated self-consistently from mass flux conservation. For this example, we assumed a coronal base pressure $p_{0,\rm base} = 0.2$ dyne cm^{-2}, which corresponds to a base density $\rho_{0,\rm base} = 4.32 \times 10^{-15}$ g cm^{-3}, consistent with observations presented in [15].

Figure 4(a) shows the outflow velocity $u_0(r)$ of the wind (black curve) and Alfvén speed $v_A(r)$ (red curve). Note that the two curves cross at $r \approx 4.5\ R_\odot$, which is the Alfvén critical point. Panel (b) shows the acceleration associated with the wave pressure force (red curve) as well as the acceleration of gravity (black curve). Note that the wave pressure force becomes dominant at $r \approx 4.5R_\odot$ because $D_{\rm wp}(r)$ exceeds gravity.

Waves are driven by imposing "footpoint"motions that are assumed to be incompressible and confined to the circular cross section of the tube. The characteristics of these motions are free parameters of the model, which we observationally constrain. In previous work [14] we have found that a root-mean-square (rms) velocity amplitude of $v_{\rm rms} \approx 40$ km s^{-1} is consistent with the observed spectral line widths and nonthermal velocities in coronal holes [9, 34, 35].

Waves launched by these footpoint motions travel upward within the flux tube, dissipate their energy over a wide range of heights, and heat the plasma to temperatures $T_0(r) \sim 1$ MK, generating pressure gradients that are an important factor in driving the solar wind. The Alfvén waves also exert a direct force on the plasma through their wave-pressure gradient, which plays a key role in producing the wind emanating from the equatorial corona.

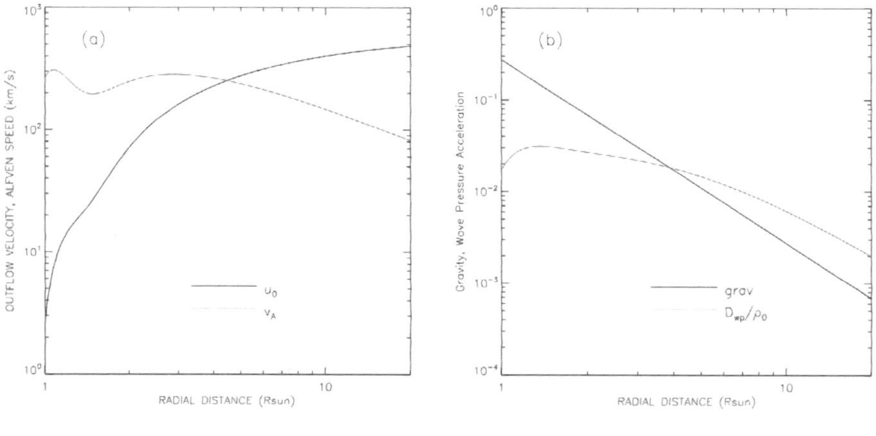

Figure 4. (a) outflow velocity (black curve) and Alfvén speed (red curve), (b) wave pressure acceleration (red curve) and gravity (black curve).

We use a spectral method of perpendicular wave numbers to describe the spatial variations of the waves perpendicular to the mean magnetic field and finite differences in the radial direction along the field. The radial grid is chosen such that for outward propagating waves the wave travel time between neighboring grid points is constant with height and equal to the simulation time step, $\Delta t = 1$ s. This allows the waves to propagate to large heights without numerical distortions. A key feature of the RMHD equations is that they retain the nonlinear terms responsible for the development of Alfvén-wave turbulence. For $v_{\rm A} > u_0$, the Alfvén waves have significant effects on the plasma as the wave turbulence causes plasma heating and the waves exert a direct force on the outflowing plasma via wave-pressure gradients.

The atmospheric model presented in Figures 3 and 4 is used as the background for our three-dimensional, time-dependent RMHD simulations of the Alfvén waves inside an open field. In the initial state there are no waves at any height. The waves are launched by imposing certain footpoint motions on the plasma at the coronal base, which is the lower boundary of the RMHD model. The simulated waves travel upward along the flux tube and partially reflect due to spatial variations in Alfvén speed $v_A(r)$ and outflow velocity $u_0(r)$. The outward propagating waves reach the outer boundary of the RMHD model ($r = 20\ R_\odot$) after about 10,891 s of simulated time. The waves are simulated for a period of 30,000 s to ensure that a statistically significant stationary state is reached. In this state there are outward and inward propagating waves at all heights, and there are waves with a broad spectrum of perpendicular wavenumbers, indicating that strong wave turbulence has developed.

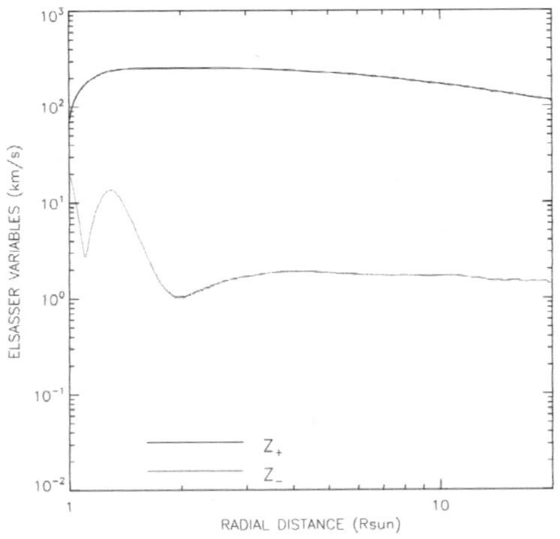

Figure 5. Elsasser variable (a) Z_+ (black) represents outward propagating waves, and (b) Z_- (blue) represents inward (reflected) waves.

The black and blue curves in Figure 5 show the rms values of the Elsasser variables. Note that the inward waves Z_- are much weaker than the outward waves. The amplitude of the inward waves $Z_-(r)$ has a sharp minimum at $r \approx 1.3\ R_\odot$. We attribute this to the fact that the Alfvén speed has a maximum near that height. The reduction in $Z_-(r)$ is due to turbulent decay of inward waves. The reduction is significant because the inward waves have a short nonlinear time scale. Therefore, the nonlinear interactions between inward and outward waves play an important role in determining the inward wave amplitudes [12].

Figure 6(a) shows the total energy density E of the simulated waves (full black curve), together with the contributions to this quantity from the kinetic energy $E_{\rm kin}$ (red curve) and magnetic energy density $E_{\rm mag}$ (green curve). Figure 6(b) shows the total energy dissipation rate $Q(r)$ of the simulated turbulence (solid black curve). This rate is the sum of the contributions from kinetic and magnetic heating rates.

4. Discussion

We demonstrated in this paper how the Alfvén waves generated in an equatorial open field creates turbulence that heats and accelerates the slow solar wind. We generated the $B(r)$ profile using magnetic field reconstructions of the observed equatorial corona. The modeled equatorial open field was considered to be the origin of the slow solar wind. The magnetic profile was used as an input for our RMHD Alfvén wave turbulence model. We excited the Alfvén waves

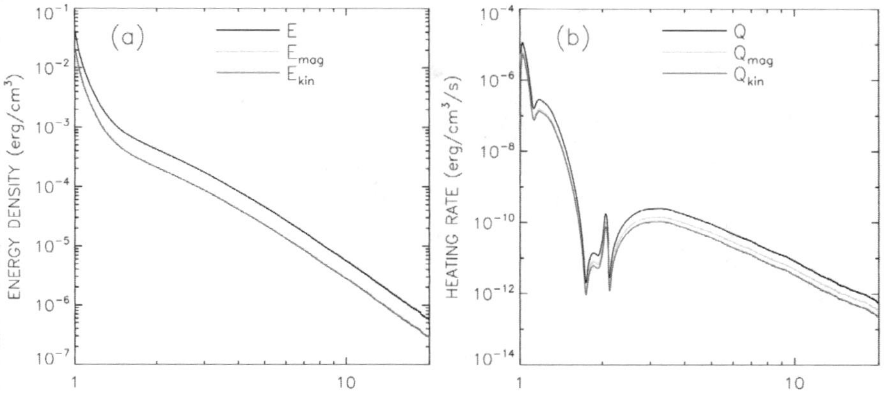

Figure 6. (a) Wave energy densities: total energy (black curve), kinetic energy (red curve), and magnetic energy (green curve). (b) Wave energy dissipation rates per unit volume: total wave dissipation rate Q (solid black curve), and contributions from $Q_{\rm kin}$ (solid red curve), and $Q_{\rm mag}$ (solid green curve).

in the model at the base of the corona and allowed them to flow through the $B(r)$ profile. We computed the expected Alfvén-wave reflection, turbulence properties, and heating rate from the base of the corona to 20 $R_\odot$.

In the case considered here, the total dissipation rate $Q_{\rm mag} + Q_{\rm kin}$ remains below the heating rate Q needed to sustain the background atmosphere. One way to increase the wave dissipation rate is to enhance the amplitude of the inward propagating waves. In our previous studies of the fast solar wind [12–14] we included density fluctuations in the Alfvén wave turbulence model and showed that density variations can, in principle, produce strong wave reflections that significantly enhance the amplitude of the inward propagating waves and thereby increase wave dissipation rate enough to heat the background atmosphere.

In our future studies of the slow solar wind, we will look for the evidence of density fluctuations from the observations. We will characterize the properties of density fluctuations from the base of the solar corona to 20 $R_\odot$, along the path of the outflowing slow solar wind using remote sensing observations and direct measurements in situ with Parker Solar Probe.

Acknowledgments
M. Asgari-Targhi is supported under contract 80NSSC23K0087 from the NASA Heliophysics Supporting Research program to the Smithsonian Astrophysical Observatory (SAO) and Columbia University.

References

[1] Zirker J B 1977 (Boulder: Colorado Assoc. Univ. Press)

[2] Parker E N 1965 *Space Sci. Rev.* **4** 666

[3] Heinemann M and Olbert S 1980 *J. Geophys. Res.* **85** 1311

[4] Velli M 1993 *Astron. & Astrophys.* **270** 304

[5] Dmitruk P, Matthaeus W H and et al L J M 2002 *Astrophys. J.* **575** 571

[6] Hollweg J V and Isenberg P A 2007 *J. Geophys. Res.* **112** 8102

[7] Suzuki T K and Inutsuka S I 2006 *J. Geophys. Res.* **111** A06101

[8] Cranmer S R, van Ballegooijen A A and Edgar R J 2007 *Astrophys. J. Suppl.* **171** 520

[9] Hahn M, Landi E and Savin D W 2012 *Astrophys. J.* **753** 36

[10] Bemporad A and Abbo L 2012 *ApJ* **751** 110

[11] Hahn M and Savin D W 2013 *ApJ* **763** 106

[12] van Ballegooijen A A and Asgari-Targhi M 2016 *Astrophys. J.* **821** 106

[13] van Ballegooijen A A and Asgari-Targhi M 2017 *Astrophys. J.* **835** 10

[14] Asgari-Targhi M, Asgari-Targhi A, Hahn M and Savin D W 2021 *Astrophys. J.* **911** 63

[15] Habbal S R, Druckmuller M and et al N A 2021 *Astrophys. J. Letters* **911** 4

[16] Geiss J, Gloeckler G and von Steiger R 1995 *Space Sci. Rev.* **72** 49–60

[17] Zhao L, Zurbuchen T H and Fisk L A 2009 *Geophys. Res. Lett.* **36** L14104

[18] Bale S D, Badman S T, Bonnell J W, Bowen T A, Burgess D, Case A W, Cattell C A, Chandran B D G, Chaston C C, Chen C H K, Drake J F, Dudock de Wit T, Eastwood J P, Ergun R E, Farrell W M, Fong C, Goetz K, Goldstein M, Goodrich K A, Harvey P R, Horbury T S, Howes G G, Kasper J C, Kellogg P J, Klimchuk J A, Korreck K E, Krasnoselskikh V V, Krucker S, Laker R, Larson D E, MacDowall R J, Maksimovic M, Malaspina D M, Martinez-Oliveros J, McComas D J, Meyer-Vernet N, Moncuquet M, Mozer F S, Phan T D, Pulupa M, Raouafi N E, Salem C, Stansby D, Stevens M, Szabo A, Velli M, Woolley T and Wygant J R 2019 *Nature* **576** 237

[19] Shebalin J V, Matthaeus W H and Montgomery D 1983 *J. Plasma Phys.* **29** 525

[20] Bhattacharjee A and Ng C S 2001 *Astrophys. J* **548** 318

[21] Howes G G and Nielson K D 2013 *J. Plasma Phys.* **20** 072302

[22] Davis C and Bewsher D 2007 *Astronomy & Geophysics* **48** 12

[23] et al H 2005 *Adv. Space Res.* **36** 1512

[24] Scherrer P H, Bogart R S and et al R I B 1995 *Solar Phys.* **162** 129

[25] van Ballegooijen A A 2004 *Astrophys. J.* **612** 519

[26] Asgari-Targhi M and van Ballegooijen A A 2012 *Astrophys. J.* **746** 81

[27] Asgari-Targhi M, van Ballegooijen A A, Cranmer S R and DeLuca E E 2013 *Astrophys. J.* **773** 111

[28] Asgari-Targhi M, Golub L, Hahn M, Karna N and Savin D W 2021 *Astrophys. J.* **910** 113

[29] Strauss H R 1976 *Phys. Fluids* **19** 134

[30] Strauss H R 1997 *J. Plasma Phys.* **57** 83

[31] Zank G P and Matthaeus W H 1992 *J. Plasma Phys.* **48** 85

[32] Verdini A and Velli M 2007 *Astrophys. J.* **662** 669

[33] Perez J C and Chandran B D G 2013 *Astrophys. J.* **776** 124

[34] Wilhelm K, Marsch E, Dwivedi B N, Hassler D M, Lemaire P, Gabriel A H and Huber M C E 1998 *The Solar Corona Above Polar Coronal Holes as Seen by SUMER on SOHO* **500** 1023

[35] Landi E and Cranmer S R 2009 *Astrophys. J.* **691** 794

Impact of Magnetic Focusing on the Transport of Energetic Electrons in the Solar Corona

Bofeng Tang[1,2], Haihong Che[1,2], Gary P. Zank[1,2]

[1] Department of Space Science, University of Alabama in Huntsville, Huntsville, AL 35899, USA

[2] Center for Space Plasma and Aeronomic Research (CSPAR), University of Alabama in Huntsville, Huntsville, AL 35805, USA

E-mail: bt0008@uah.edu, Haihong.Che@uah.edu,garyp.zank@gmail.com

Abstract. Observations of Type III radio bursts discovered that electron beams with power-law energy spectra are commonly produced during solar flares. The locations of these electron beams are $\sim$ 300 Mm above the particle acceleration region of the photosphere, and the velocities range from 3 to 10 times the local background electron thermal velocity. However, the mechanism that can commonly produce electron beams during the propagation of energetic electrons with power-law energy spectra in the corona remains unclear. In this paper, using kinetic transport theory, we find for the first time that the magnetic focusing effect governs the formation of electron beams over the observational desired distance in the corona. The magnetic focusing effect can sharply increase the bulk velocity of energetic electrons to the observed electron beam velocity within 0.4 solar radii (300 Mm) as they escape from the acceleration region and propagate upward along magnetic field lines. In more rapidly decreasing magnetic fields, energetic electrons with a harder power-law energy spectrum can generate a higher bulk velocity, producing type III radio bursts at a location much closer to the acceleration region. During propagation, the spectral index of the energetic electrons is unchanged.

1. Introduction

Solar type III radio bursts, first discovered by Wild & McCready [1], are transient radio emissions drifting from higher frequencies to lower frequencies as a function of time. The type III radio bursts are commonly believed to be produced by propagating electron beams via an electron two-stream instability (ETSI) [2,3] as was proposed by Ginzburg & Zhelezniakov (1958) [4]. Recently discovered radio bursts showed that electrons beams are commonly produced in solar flares [5–7]. In particular, numerous nanoflare-associated radio bursts were discovered [8], implying that electron beams are distributed widely in the solar corona, and very probably contribute to the formation of the electron halo in the solar wind [9, 10]. However, the previous studies [5, 11] focus on the maintenance of the electron beams after triggering the Type III radio bursts rather than the formation of the beams before the triggering, and the common origin of the electron beams is poorly understood.

It is well known that solar flares accelerate electrons to a power-law energy distribution [12–14] in the magnetic reconnection (MR) region. Theory and particle-in-cell simulations show that the power-law energetic electrons can be produced in the diffusion region of MR [15–17]. However, the bulk velocity of the energetic electrons does not usually exceed the thermal velocity of the electrons due to intense plasma heating in the acceleration region. As a result, energetic electrons cannot trigger the electron two-stream instability (ETSI) immediately and hence do not produce Type III radio bursts in the acceleration region [3,17]. Observations also found that as the energetic electrons continue propagating a certain distance away from the acceleration region, Type III radio bursts are produced [5,6,18–21]. Hence the height of the MR acceleration region where energetic electrons are produced initially is lower than the site where the Type III radio bursts are excited, suggesting that energetic electrons with power-law energy spectra develop into electron beams after they escape from the acceleration region and propagate up along magnetic field lines.

What effects commonly govern the electron beam formation in the solar corona is unclear. The electron Coulomb collision rate in the inner corona is $\nu_e \sim 10^{-6} n_e T_e \sim 10^{-4}$ s^{-1} and the Coulomb heating timescale is about 10^4 s. During this period, keV electrons (with a typical velocity about 1×10^3 km/s) travels 100 solar radii. Thus, the impact of Coulomb collisions on the transport of energetic electrons is negligible [22, 23]. Turbulent acceleration is another significant effect [5], but our knowledge about solar corona turbulence is poor, and its origin and intensity are uncertain. One certain effect that can affect electrons along the magnetic field is the magnetic field gradient. The open magnetic field lines forming during magnetic reconnection are highly curved and decrease as they extend from the lower corona to space [24]. Since the magnetic fields of magnetic loops change slowly in time and space compared to the time scale of electrons traveling in the corona, the magnetic moment of the electrons is approximately conserved [25]. Consequently, the conservation of electron magnetic momentum leads to a increased parallel velocity and a decreased perpendicular velocity simultaneously, dubbed as the magnetic focusing effect. Magnetic focusing is an intrinsic feature of the solar magnetic field and can impact the transport of particles throughout the whole solar corona and solar wind. It has been well known that the magnetic focusing effect plays an decisive role in the formation of the solar wind electron strahl at 1 AU, but its role in the development of the fast electron beams within a far-more shorter distance (~ 0.002 AU) in the corona has not been well studied.

In this paper, using kinetic transport theory, we show that magnetic focusing effectively transfers the perpendicular momentum to the forward parallel momentum when energetic electrons run away from the acceleration region of solar flares and move up along magnetic field lines. For the first time, we find that the bulk velocity of energetic electrons (electron beam) increases rapidly over a small distance of less than 0.4 solar radii (300 Mm) to much higher than the local electron thermal velocity, indicating that an ETSI can be triggered and a Type III radio burst produced. This result is consistent with observations that observations that the electron beams that produce type III radio bursts are usually found above 100 Mm above photosphere. In a more rapidly decreasing magnetic field, energetic electrons with a harder spectra can develop a faster beam, which triggers the ETSI and produce type III radio bursts at a location much closer to the acceleration region. The spectral index of the power-law energy distribution remains the same as that initially inside the acceleration region.

2. Theoretical model

We modify the gyrophase averaged kinetic transport equation for a gyrotropic electron distribution function $f(\boldsymbol{x}, v, \mu, t)$ [26] to include drift along a general form magnetic field:

$$\begin{aligned}&\frac{\partial f}{\partial t}+(v\mu\boldsymbol{b}+\boldsymbol{U})\cdot\nabla f\\&+\left[\left(-\frac{\partial\boldsymbol{U}}{\partial t}-\boldsymbol{U}\cdot\nabla\boldsymbol{U}\right)\cdot\boldsymbol{b}\frac{\mu}{v}-\frac{1-\mu^2}{2}\nabla\cdot\boldsymbol{U}-\frac{3\mu^2-1}{2}\boldsymbol{b}\cdot\nabla\boldsymbol{U}\cdot\boldsymbol{b}+\frac{qE_{\|}\mu}{mv}\right]v\frac{\partial f}{\partial v}\\&+\frac{1-\mu^2}{2}\left[v\nabla\cdot\boldsymbol{b}+\mu\nabla\cdot\boldsymbol{U}-\left(\frac{\partial\boldsymbol{U}}{\partial t}+\boldsymbol{U}\cdot\nabla\boldsymbol{U}\right)\cdot\boldsymbol{b}\frac{2}{v}-3\mu\boldsymbol{b}\cdot\nabla\boldsymbol{U}\cdot\boldsymbol{b}+\frac{2qE_{\|}}{mv}\right]\frac{\partial f}{\partial\mu}\\&=\left(\frac{\delta f}{\delta t}\right)_{\rm sc}\end{aligned}\tag{1}$$

where $\boldsymbol{x}$ is the position in the rest frame, (v,μ) the velocity magnitude and the cosine of pitch angle measured in a moving frame drifting with $\boldsymbol{U}=c(\boldsymbol{E}\times\boldsymbol{B})/B^2+U_{\|}\boldsymbol{b}$ with respect to the rest frame of the Sun, where $U_{\|}=\boldsymbol{U}\cdot\boldsymbol{b}$, $\boldsymbol{E}$ and $\boldsymbol{B}$ are the external fields, and $\boldsymbol{b}\equiv\boldsymbol{B}/B$ the unit vector of along the magnetic field. $E_{\|}$ and $(\delta f/\delta t)_{\rm sc}$ are the parallel electric field and scattering mechanisms, and both are ignored in the present paper since magnetic focusing alone is considered.

Consider a constant drifting velocity of the moving frame along the magnetic field $\boldsymbol{U}=U\boldsymbol{b}$, it is easy to show that $\nabla\cdot\boldsymbol{U}=U\nabla\cdot\boldsymbol{b}$. The derivative divergence of $\boldsymbol{b}$ can be expressed in terms of the magnetic field $\boldsymbol{B}$:

$$\nabla\cdot\boldsymbol{b}=-\frac{\boldsymbol{B}\cdot\nabla B}{B^2}.\tag{2}$$

where $\nabla\cdot\boldsymbol{B}=0$ has been taken into consideration. In a global spherical coordinate system whose origin is the center of the Sun, let $\mathbf{B}=B_r\hat{e}_r+B_\theta\hat{e}_\theta+B_\phi\hat{e}_\phi$, where $\hat{e}_r$, $\hat{e}_\theta$ and $\hat{e}_\phi$ are the radial, polar and azimuthal unit vectors, respectively. We are interested in the outward transport of energetic electrons in solar flares. For the solar flare background, the magnetic loops are approximately force-free, and the magnetic field model proposed by Antiochos et al. [24] is commonly accepted. In this model, it is assumed that the radial component is much larger than the other two components, i.e., $B\approx B_r$ [24]. The divergence of $\boldsymbol{b}$ becomes:

$$\nabla\cdot\boldsymbol{b}\approx-\frac{\partial\ln B}{\partial r},\tag{3}$$

where r is the radial distance. The velocity v of electrons is always much larger than the plasma background velocity U, i.e., $U\ll v$, making it safe to ignore U. Therefore, the transport equation (1) reduces to:

$$\frac{\partial f}{\partial t}+\mu v\frac{\partial f}{\partial r}-\frac{1}{2}\frac{d\ln B}{dr}(1-\mu^2)v\frac{\partial f}{\partial\mu}=0.\tag{4}$$

Compared with equation (1), the magnetic field in equation (4) is restricted to a magnetic field configuration whose non-radial components are much weaker than radial. As a result, the effect of focusing by the non-radial components of the magnetic field is negligible, and thus the distribution function f is nearly independent of θ and ϕ in the global spherical coordinate system. Therefore, in the following, we only consider the radial magnetic field.

The magnetic field of a solar flare varies with time, and its timescale is much larger than that of energetic electrons transport in the solar corona. For example, a 10 keV electron whose velocity is about 10^4 km/s needs 10 s to travel one solar radius, which is much shorter than the 1000 s time scale of a typical solar flare [6]. Hence it is reasonable to assume that energetic electrons experience an approximately "stationary" magnetic field as they transport in the solar corona—this is the condition for the conservation of magnetic moment, which causes magnetic focusing in a spatially decreasing magnetic field. As a magnetic field decreases with radial distance r, the simplest case is $B = B_0/r^2$, which ensures the divergence-free condition of magnetic field, otherwise, the magnetic field must include the other two components but they are assumed to be sufficiently small compared to the radial component and can be safely ignored. To choose the radial magnetic field configuration, we follow the reference scenario in the modeling of solar magnetic field by Banaszkiewicz et al. [27] and open magnetic field lines of solar flares by Antiochos et al. [24]. We approximate the outward magnetic field lines of the solar flares as $B = B_0/r^{n_\mathrm{B}}$, where B_0 is the magnetic field at the inner boundary, $n_\mathrm{B} = 1, 2, 3, \cdots$. A more general form of the magnetic field can be expressed as a summation of a series of $1/r^{n_\mathrm{B}}$. With this configuration, we can understand how the spatial evolution of the magnetic field impacts magnetic focusing and the resulting beam development. Hence, $d \ln B/dr = -n_\mathrm{B}/r$ and only the value of n_B enters in the transport equation (4) while the exact value of the magnetic field disappears:

$$\frac{\partial f}{\partial t} + \mu v \frac{\partial f}{\partial r} + \frac{n_\mathrm{B}}{2} \frac{(1-\mu^2)v}{r} \frac{\partial f}{\partial \mu} = 0. \tag{5}$$

The energetic electron VDF produced at the acceleration region of MR is used as the initialization at the inner boundary. A combination of a drifting Maxwellian (f_m) and a drifting power-law (f_p) is chosen to mimic the energetic electron VDFs with a drifting velocity U_d. Hence,

$$f(v, \mu \geq 0) = \delta f_\mathrm{m}(v,\mu) + (1-\delta) f_\mathrm{p}(v,\mu), \tag{6}$$

$$f_\mathrm{m}(v,\mu) = n_0 \left(\frac{m}{2\pi k_\mathrm{B} T_\mathrm{c0}}\right)^{3/2} \exp\left\{-\frac{m}{2k_\mathrm{B}T_\mathrm{c0}} \left[(v\mu - U_\mathrm{d})^2 + v^2\left(1-\mu^2\right)\right]\right\}, \tag{7}$$

$$f_\mathrm{p}(v,\mu) = n_0 \frac{1}{4\pi} \frac{\alpha - 3}{v_\mathrm{c}^{3-\alpha}} \left\{\left[(v\mu - U_\mathrm{d})^2 + v^2(1-\mu^2)\right]^{1/2}\right\}^{-\alpha}, \quad (v\mu - U_\mathrm{d})^2 + v^2(1-\mu^2) \geq v_\mathrm{c}^2. \tag{8}$$

where δ is the weight of the Maxwellian component f_m, T_c0 the temperature of the Maxwellian component, n_0 the total electron number density at the inner boundary, v_c the critical velocity separating the power-law component (energetic electrons) from the Maxwellian, and α the spectral index of the power-law component. The upper-left panel in Figure 1 shows an example of the electron VDF at the inner boundary in the $(v_\parallel, v_\perp)$ plane in the case of $\delta = 0.8$, $\alpha = 4$, $U_\mathrm{d} = 1 v_\mathrm{th0}$ and $n_\mathrm{B} = 2$.

We numerically solve the transport equation (5) with boundary condition (6) – (8) to study the transport of energetic electrons in the corona from 1.05 to 3.05 R_s, where R_s is the solar radius. The inner boundary of 1.05 R_s is chosen by assuming the accelerated electrons are injected in the acceleration region of MR, which is in the lower corona [5, 19, 28], and the outer boundary of 3.05 R_s is the typical size of magnetic loops [24]. We choose the critical velocity $v_\mathrm{c} = 1.5 v_\mathrm{th0}$, where $v_\mathrm{th0} = (2k_B T_\mathrm{c0}/m)^{1/2}$ is the electron thermal velocity at the inner boundary. We adjust the parameters α and U_d to mimic the diverse power-law electron energy spectra and the drifts of electron outflows produced by MR acceleration. We adjust n_B to mimic the open field lines in MR produced by magnetic field lines with different curvatures. The various combinations of different α, U_d and n_B used in the calculation are listed in Table 1 and the meaning will be explained and discussed in detail in the following sections. The numerical stationary solution is obtained after the code approaches the steady final state.

Table 1. Combinations of different Parameters

α	$U_{\rm d}(v_{\rm th0})$	$n_{\rm B}$
4	0	3
4	1	1
4	1	2
4	1	3
4	1	5
4	0.5	3
6	0.5	3
8	0.5	3

3. Numerical results

Magnetic focusing is the direct consequence of the conservation of magnetic moment defined as $\mu_M = mv_\perp^2/2B$. The magnetic field decreases as the electrons move outward along open magnetic field lines. The conservation of μ_M results in a decrease of the perpendicular electron velocity, and conservation of electron kinetic energy increases the parallel electron velocity.

The steady analytical solution $f(r, v, \mu)$ of equation (5) can be obtained by the method of characteristics [29, 30]:

$$f(r,v,\mu) = \begin{cases} f_0(r_0,v,\mu), & \sqrt{1-\left(\dfrac{r_0}{r}\right)^{n_{\rm B}}} < \mu < 1 \\ 0. & -1 < \mu < \sqrt{1-\left(\dfrac{r_0}{r}\right)^{n_{\rm B}}} \end{cases}, \tag{9}$$

where $f_0(r_0, v, \mu)$ is any given inner boundary condition r_0. Figure 1 shows the numerical solution for the radial evolution of the electron VDF at 1.05, 1.5, 2 and 3 $R_{\rm s}$ respectively, for a velocity distribution with $\delta = 0.8$, $\alpha = 4$, $U_{\rm d} = 1v_{\rm th0}$ and $n_{\rm B} = 2$ at the inner boundary. As distance increases, the electron VDFs become increasing anisotropic, i.e., the maximum pitch-angle of the electron VDF decreases with the increasing distance, which is due to focusing induced by a decreasing magnetic field, the maximum value of the electron perpendicular velocity decreases while that of the parallel velocity increases, leading to the formation of a group of fast outward-moving electrons - an electron beam.

To study the radial evolution of the bulk velocity of energetic electrons, we define it as:

$$u_{\rm b} = \frac{2\pi \displaystyle\int_{v_{\rm min}(r,\mu)}^{\infty} dv \int_{\mu_{\rm min}(r)}^{1} d\mu v^3 \mu f}{2\pi \displaystyle\int_{v_{\rm min}(r,\mu)}^{\infty} dv \int_{\mu_{\rm min}(r)}^{1} d\mu v^2 f}. \tag{10}$$

This definition is similar to the definition of the mean electron velocity within the beam [7]. Hence, $\mu_{\rm min}(r) = \sqrt{1-(r_0/r)^{n_{\rm B}}}$ is the cosine of the maximum pitch-angle of the electron VDF at distance r, which is given by the analytical solution (9). At r_0, $\mu_{\rm min} = 0$, and increases with increasing distance and $v_{\rm min} = U_{\rm d}\mu + \sqrt{U_{\rm d}^2(\mu^2-1) + v_{\rm c}^2}$ is the critical velocity of the power-law

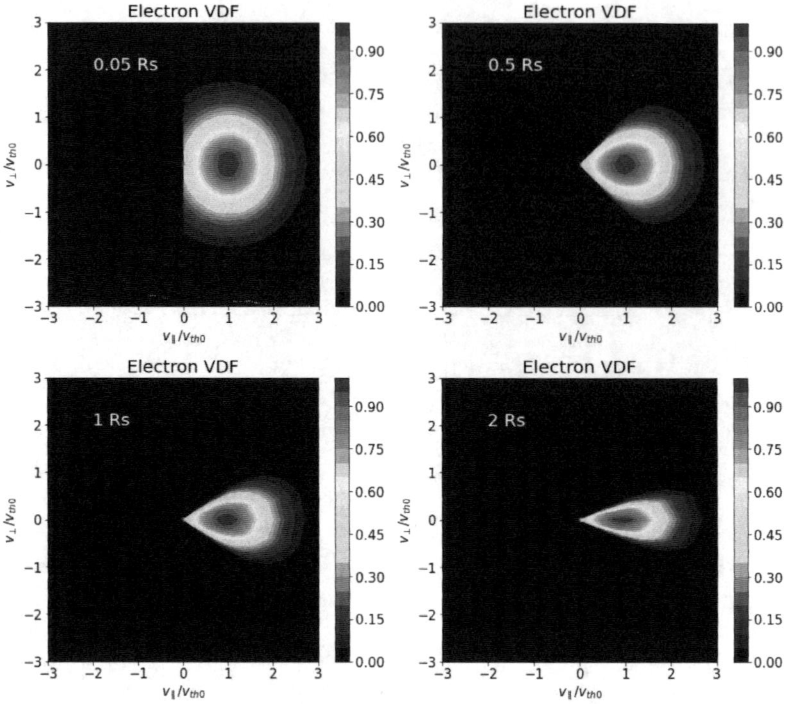

Figure 1. 2D contour plots of the electron VDF in the moving frame at 1.05, 1.5, 2 and 3 R_s for $U_d = 1v_{th0}$ and $B = B_0/r^2$.

component, for which $v_{min} = v_c$ if $U_d = 0$. In the case of $U_d \neq 0$, $f(r, v, \mu)$ and v_{min} are highly anisotropic, and the analytical solution of u_b is hard to obtain.

The radial evolution of the bulk velocity of the energetic electron u_b in the rest frame of the Sun is shown in Figure 2, A1 and A2. Figure 2 A1 has a Maxwellian velocity distribution with weight $\delta = 1$ at the inner boundary. In Figure 2 A2 the electron VDF at the inner boundary is a Maxwellian with weight $\delta = 0.8$ plus a power-law velocity tail (energetic electrons) with index $\alpha = 4$ and weight 0.2 $(= 1 - \delta)$. The index $\alpha = 4$ is estimated according to the lower bound of the observable index of electron energy spectra which is about 3 [6,18,31,32], and the weight 0.2 $(= 1 - \delta)$ is estimated by the typical acceleration efficiency found in PIC simulations [16,17]. Different velocity starting points of u_b are caused by the different drifting velocities U_d (0, $0.5v_{th0}$ and v_{th0}) of the electron VDF at the inner boundary.

It can be seen that u_b under all conditions increases sharply from the inner boundary 1.05 to 1.4 R_s (i.e., over a distance of 0.4 R_s, or 300 Mm), indicating that magnetic focusing increases the parallel bluk velocity when energetic electrons move upward along magnetic field lines. Comparing with A1, Figure 2 A2 shows that the energetic electrons with a power-law energy spectrum escaping from the acceleration region can be more efficiently focused to reach a higher bulk velocity, which is about three times larger than the thermal velocity v_{th0}. The decreasing electron temperature with distance in the corona leads to a decrease of the local electron thermal velocity $(v_{th0} > v_{th})$. Therefore above a distance of 0.4 R_s (300 Mm), energetic electrons likely have a bulk velocity u_b that is a much larger value than the local electron thermal velocity, i.e. $u_b \gg v_{th}$. These energetic electrons with large bulk velocity can trigger a strong ETSI and produce Type III radio bursts [3,33], which is consistent with the observations that type III radio bursts are always found above a certain distance (> 300 Mm) from the acceleration

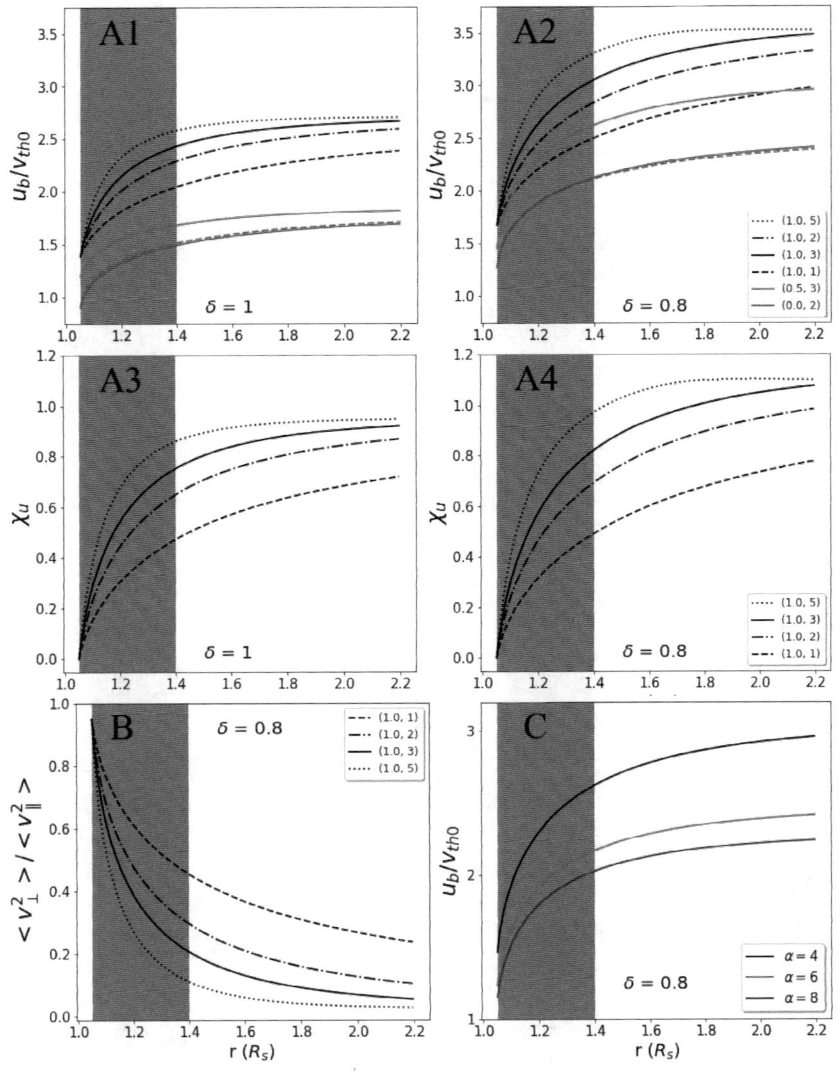

Figure 2. (A1, A2): Radial evolution of the electron beam velocity u_b in the rest frame. A1: The inner boundary condition of the electron VDF is an isotropic Maxwellian only. A2: The inner boundary condition of the electron VDF is an isotropic Maxwellian with weight $\delta = 0.8$ plus an isotropic power-law with $\alpha = 4$ and weight 0.2 ($= 1 - \delta$). The drifting velocity U_d and n_B of magnetic field $B = B_0/r^{n_B}$ are shown in the brackets (U_d, n_B). (A3, A4): Radial evolution of electron beam velocity growth rate χ_u corresponding to A1 and A2 respectively. B: Radial evolution of $< v_\perp^2 > / < v_\parallel^2 >$ for an electron beam. Line colors and styles have the same meaning as in A2. C: Radial revolution of u_b with different power-law indices $\alpha = 4$, 6 and 8. The shaded areas indicate the distance range from 1.05 to 1.4 R_s.

region [19, 20]. We find that even though the hot electrons do not develop outflow in the acceleration region, the energetic electrons can escape and generate a bulk flow with velocity u_b much larger than the electron thermal velocity v_{th0}, and trigger ETSI.

The differences between Figure 2 A2 and A1 illustrate how the magnetic field and power-law tail affect the increase in the bulk velocity of the energetic electrons. First, the bulk velocity

u_b in A2 has a higher initial value at the inner boundary and a higher final value at the outer boundary than those in A1. The power-law tail in A2 provides more energetic electrons, leading to a higher initial u_b. To study the increased efficiency, we define a growth rate for the electron bulk velocity as $\chi_u = (u_\mathrm{b} - u_\mathrm{b0})/u_\mathrm{b0}$ where u_b0 is the initial u_b at the inner boundary. The radial evolution of the growth rates χ_u are shown in A3 and A4 in Figure 2, where A3 is derived from A1 and A4 from A2. The growth rates χ_u in A4 are larger than those in A3, showing that magnetic focusing of electrons with a power-law tail leads to a higher bulk velocity than the case of a Maxwellian distribution only. Therefore magnetic focusing can increase the electron bulk velocity much more rapidly if more energetic electrons with a power-law tail are present, thus triggering ETSI and producing type III radio bursts. Second, the n_B in magnetic field $B = B_0/r^{n_\mathrm{B}}$ also affects the radial evolution of u_b and its growth rate χ_u. As n_B increases from $n_\mathrm{B} = 1$ to 3, and to 5, both the electron bulk velocity u_b and its growth rate χ_u increase faster and reach a larger value. Consequently, in a more rapidly decreasing magnetic field, we expect more Type III radio bursts to occur at distances closer to the acceleration region of MR.

Magnetic focusing causes the parallel acceleration of electrons along the magnetic field line by transferring electron kinetic energy from the perpendicular direction to the parallel direction. We calculate the ratio of the perpendicular kinetic energy to the parallel kinetic energy of electrons ($< v_\perp^2 > / < v_\parallel^2 >$) using

$$\frac{< v_\perp^2 >}{< v_\parallel^2 >} = \frac{\int_0^\infty v_\perp^2 f d\boldsymbol{v}}{\int_0^\infty v_\parallel^2 f d\boldsymbol{v}}. \tag{11}$$

Figure 2B displays the radial evolution of the kinetic energy ratio in the A2 case with $\delta = 0.8$ and $U_\mathrm{d} = v_\mathrm{th0}$. We find that the kinetic energy ratios start from less than 1 and decrease with distance. For an isotropic electron VDF at the inner boundary, the kinetic energy ratio is 2 since the isotropic VDF has one degree of freedom in the parallel direction and two in the perpendicular direction. However, in our case, the electron VDF with a drifting velocity $U_\mathrm{d} \neq 0$ is anisotropic, and its kinetic energy ratio is less than 2. The radial evolution of the kinetic energy ratio decreases with distance which contrasts with the radial evolution of the electron bulk velocity u_b and its growth rate χ_u. It is noticeable that the kinetic energy ratio decreases sharply in the same distance range from 1.05 to 1.4 R_s, while u_b and χ_u, on the contrary, increase sharply. The radial evolution of kinetic energy ratio caused by the magnetic focusing effect is dominated by the value of n_B in $B = B_0/r^{n_\mathrm{B}}$: the larger n_B, the faster the decrease of kinetic energy ratio.

Panel C of Figure 2 shows the radial evolution of electron beam velocity u_b of electrons with different energy spectral indices α. The electron bulk speed u_b with a smaller spectral index α has a higher initial value since the electron VDF with a smaller spectral index α possesses more energetic electrons. The smaller spectral index α leads to a faster increase of u_b with radial distance. Similarly, the maximum increase is in the same gray-shaded area.

We find that the electron energy spectral index α of energetic electrons is preserved during the upward propagation in the presence of magnetic focusing. We plot a parallel cut of the electron VDF ($v_\parallel > 0$) in Figure 3 and consider the radial evolution of the energy spectra index α of the power-law tail. The energy spectral index α does not change with distance, and the more distant electron VDF in interplanetary space has the same index α as the electron VDF at the inner boundary. Consequently, the information embedded in the distribution function of the energetic electrons produced in the acceleration region of the solar flare is preserved.

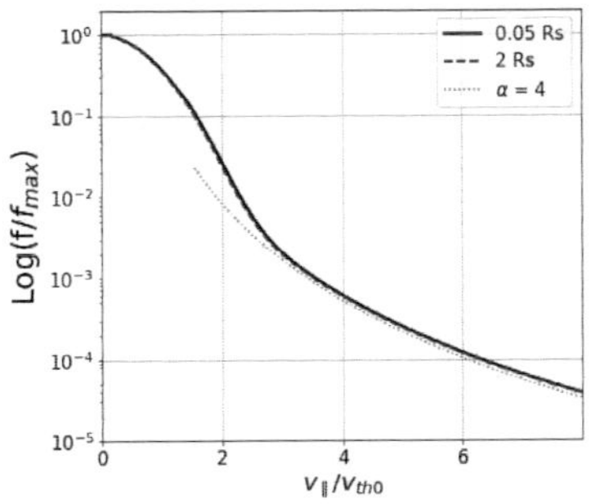

Figure 3. A parallel cut of the electron VDFs at the inner and outer boundary. The normalized electron VDF is calculated as $f/f_{\max}$ where $f_{\max}$ is the maximum. At $v_\parallel > 0$, the outwardly propagating electron VDFs overlap and both have $\alpha = 4$.

4. Discussion and conclusion

Observations of coronal Type III radio bursts discovered that electron beams are commonly produced in solar flares [6, 8, 34], and the location of these beams is beyond some 300 Mm from the acceleration region. However, the origin of electron beams is poorly understood. In this paper, we investigate the transport of energetic electrons produced in the acceleration region of solar flares as they propagate upward along the magnetic field lines of a solar flare. Magnetic focusing is an intrinsic property of a decreasing strength magnetic field, typical of the solar corona and solar wind, which transfers electron perpendicular momentum and kinetic energy to the parallel direction. We ignore Coulomb and wave-particle scattering terms and construct a kinetic transport equation for energetic electrons in the solar corona subject to magnetic focusing. By numerically solving the kinetic transport equation, we investigate how magnetic focusing affects the transport of energetic electrons along outward magnetic field lines in the solar corona. Observations established that electrons are commonly accelerated by solar flares to power-law energy spectra in the acceleration region. Here we mimic the acceleration process by choosing the initial eVDF at the inner boundary with a power-law tail, and the power-law index range varies from 4 to 8, consistent with the observations.

We find that the magnetic focusing effect sharply increases the bulk velocity of energetic electrons over an observational desired distance of less than 0.4 solar radii (300 Mm) to be much larger than the local electron thermal velocity. Energetic electrons with bulk velocity much larger than the local electron thermal velocity can trigger type III radio bursts. This is consistent with observations that the electron beams that produce type III radio bursts are usually found above 300 Mm above photosphere. In a more rapidly decreasing magnetic field (larger $n_{\rm B}$), energetic electrons with a harder power-law energy spectrum (smaller α) can drive a faster beam. Energetic electrons with higher bulk velocity can trigger the ETSI and produce type III radio bursts at a location much closer to the acceleration region. Magnetic focusing does not change the initial spectral index α of energetic electrons as they propagate outward along magnetic field lines. Thus, outwardly moving energetic electrons preserve the electron distribution information from the region where they initially formed.

This paper focuses on the magnetic focusing effect. It will be necessary to extend our numerical model to include Coulomb collisions and turbulence, investigating the competition between these three effects in forming electron beams. This will be presented in a following paper.

Acknowledgment. B. T. and H. C. acknowledge partial support by NASA HSR program No. 80NSSC21K0031, a Heliophysics Career award No. 80NSSC19K1106, and NSF CAREER 2144324. The authors acknowledge the partial support of an NSF EPSCoR RII-Track-1

Cooperative Agreement OIA-2148653.

References

[1] Wild J and McCready L 1950 *Australian Journal of Chemistry* **3** 387–398
[2] Aschwanden M J 2002 *Particle acceleration and kinematics in solar flares* 1–227
[3] Che H 2016 *Modern Physics Letters A* **31** 1630018
[4] Ginzburg V and Zhelezniakov V 1958 *Soviet Astronomy* **2** 653
[5] Reid H A, Vilmer N and Kontar E P 2011 *Astronomy & Astrophysics* **529** A66
[6] Benz A O 2017 *Living Reviews in Solar Physics* **14** 1–59
[7] Reid H A and Kontar E P 2018 *The Astrophysical Journal* **867** 158
[8] Saint-Hilaire P, Vilmer N and Kerdraon A 2012 *The Astrophysical Journal* **762** 60
[9] Che H and Goldstein M 2014 *The Astrophysical Journal Letters* **795** L38
[10] Che H, Goldstein M, Salem C and Viñas A 2019 *The Astrophysical Journal* **883** 151
[11] Li B, Robinson P A and Cairns I H 2006 *Physical Review Letters* **96** 145005
[12] Krucker S, Oakley P and Lin R 2009 *The Astrophysical Journal* **691** 806
[13] Lin R 2011 *Space science reviews* **159** 421
[14] Zharkova V V, Arzner K, Benz A O, Browning P, Dauphin C, Emslie A G, Fletcher L, Kontar E P, Mann G, Onofri M *et al.* 2011 *Space science reviews* **159** 357
[15] Zank G, Le Roux J, Webb G, Dosch A and Khabarova O 2014 *The Astrophysical Journal* **797** 28
[16] Che H and Zank G 2020 *The Astrophysical Journal* **889** 11
[17] Che H, Zank G, Benz A, Tang B and Crawford C 2021 *The Astrophysical Journal* **908** 72
[18] Krucker S, Kontar E, Christe S and Lin R 2007 *The Astrophysical Journal Letters* **663** L109
[19] Chen B, Bastian T S, White S M, Gary D E, Perley R, Rupen M and Carlson B 2013 *The Astrophysical Journal Letters* **763** L21
[20] Chen B, Yu S, Battaglia M, Farid S, Savcheva A, Reeves K K, Krucker S, Bastian T, Guo F and Tassev S 2018 *The Astrophysical Journal* **866** 62
[21] Benz A O 2012 *Plasma astrophysics: Kinetic processes in solar and stellar coronae* vol 184 (Springer Science & Business Media)
[22] Fuchs V, Cairns R, Lashmore-Davies C and Shoucri M 1986 *The Physics of fluids* **29** 2931–2936
[23] Berčič L, Landi S and Maksimović M 2021 *Journal of Geophysical Research: Space Physics* **126** e2020JA028864
[24] Antiochos S, DeVore C, Karpen J and Mikić Z 2007 *The Astrophysical Journal* **671** 936
[25] Marsch E 2006 *Living Reviews in Solar Physics* **3** 1
[26] Zank G P 2014 *Transport Processes in Space Physics and Astrophysics* (*Lecture Notes in Physics, Berlin Springer Verlag* vol 877) (Springer-Verlag)
[27] Banaszkiewicz M, Axford W and McKenzie J 1998 *Astronomy and Astrophysics* **337** 940–944
[28] Cairns I H, Lobzin V, Donea A, Tingay S, McCauley P, Oberoi D, Duffin R, Reiner M, Hurley-Walker N, Kudryavtseva N *et al.* 2018 *Scientific reports* **8** 1–12
[29] Sone Y and Sugimoto H 1993 *Physics of Fluids A: Fluid Dynamics* **5** 1491–1511
[30] Tang B, Zank G P and Kolobov V I 2020 *The Astrophysical Journal* **892** 95
[31] Lin R, Evans L and Fainberg J 1973 *Astrophysical Letters* **14** 191
[32] Lin R, Potter D, Gurnett D and Scarf F 1981 *The Astrophysical Journal* **251** 364–373
[33] Che H, Goldstein M L, Diamond P H and Sagdeev R Z 2017 *Proceedings of the National Academy of Sciences* **114** 1502–1507
[34] Reid H A S and Ratcliffe H 2014 *Research in Astronomy and Astrophysics* **14** 773

Journal of Physics: Conference Series **2544** (2023) 012005 doi:10.1088/1742-6596/2544/1/012005

Anomalous Transport of Energetic Particles Interacting with Dynamic Small-Scale Magnetic Flux Rope Structures

Jakobus A le Roux[1,2]

[1] Department of Space Science, University of Alabama in Huntsville (UAH), 320 Sparkman Drive, AL 35805, USA

[2]Center of Space Plasma Science and Aeronomic Research (CSPAR), University of Alabama in Huntsville (UAH), 320 Sparkman Drive, AL 35805, USA

jar0013@uah.edu

Abstract. le Roux and Zank [25] showed previously how one can derive from first principles a pitch-angle dependent fractional diffusion-advection kinetic equation to model the anomalous diffusion of energetic particles interacting with small-scale magnetic flux ropes (SMFRs) in the inner heliosphere on the basis of the standard focused transport equation. This equation has the following limitations: (1) The asymptotic power law of a Lévy distribution was specified to model the non-Gaussian statistics of the disturbed energetic particle trajectories generated during energetic particle interaction with numerous SMFRs. The second moment (variance) and higher moments of the Lévy distribution are infinite, indicating over-efficient non-local transport that is scale-free. (2) The theory does not naturally allow for a transition of anomalous transport to normal diffusion, or to a different anomalous diffusion state. An outline of a derivation is presented in which an exponentially truncated Lévy distribution was specified instead, resulting in a tempered fractional diffusion-advection kinetic equation that addresses these two concerns.

1. Introduction

Nearly incompressible MHD turbulence theory [1, 2, 3], MHD simulations [4], and inner heliospheric observations [5, 6] suggested for a long time the existence of a dominant, non-propagating quasi two-dimensional (2D) turbulence component in the solar wind containing coherent small-scale magnetic flux rope (SMFR) structures. In support of this view, an unprecedented number of SMFRs were identified more recently in the inner heliosphere with especially the Grad-Shafranov method [7], but also using the wavelet analysis and the Magnetic Helicity-PVI method [8, 9].

Recent evidence of solar energetic particles (SEPs) being trapped between the boundaries of two neighboring SMFRs was found close to the Sun in Parker Solar Probe data during Orbit 5 between ~0.2-0.45 AU [9]. SEP trapping by SMFRs was reported previously [10, 11]. This suggests that suprathermal particle propagation through a collection of SMFRs in the perpendicular direction should statistically be subdiffusive. Results of simulations indicate systematic energy gain of energetic particles while being trapped in contracting magnetic islands [12], suggesting that energy gain could potentially be superdiffusive statistically if suprathermal particles interact with numerous dynamic SMFR structures. Furthermore, 2D MHD turbulence simulations featuring a sea of magnetic islands perpendicular to the mean magnetic/guide field generate a probability distribution function (PDF) of the perpendicular magnetic field component increments that is statistically non-Gaussian with prominent power-law tails produced by intermittent large field component increments at island boundaries [13]. The PDF of the simulation by Greco et al. reproduced quite well the PDF of ACE spacecraft observations of perpendicular field component increments at 1 AU [13]. The non-Gaussian PDF of out-of-plane current densities of SMFRs identified in Wind data and reconstructed with the Grad-Shafranov method also agrees with the PDF of the current densities of the 2D MHD turbulence simulation of Greco et al.– see [14]. These PDFs provide additional support for a strong presence of SMFRs separated by small-scale current sheets in the inner heliosphere, and imply that suprathermal particles following the magnetic field lines of SMFR structures in the solar wind should exhibit disturbed guiding center trajectories obeying non-Gaussian statistics. Thus, there is good reason to think that the transport of energetic particles propagating through solar wind regions with numerous SMFRs is anomalous. This is confirmed by a number of publications that reported superdiffusive spatial and momentum transport of energetic particles when particle trajectories are traced in MHD turbulence simulation results containing SMFRs separated by small-scale current sheets [15, 16, 17, 18].

However, anomalous transport can be viewed as an intermediate time-scale phenomenon, because in complex biological, financial, neutral and plasma fluid dynamical systems, anomalous diffusion crosses over into normal diffusion after a certain correlation time [19, 20, 21]. Examples of such transitions in the inner heliosphere are: (i) SEPs escaping from solar flare sites arriving early at 1 AU can be thought of as superdiffusive, while those that arrive later are more diffusive [22], and (ii) simulations of energetic particle propagation in turbulence where the 2D turbulence component dominates the slab component exhibits a transition of perpendicular subdiffusion to normal perpendicular diffusion [23]. To this it has to be added that there are reports of superdiffusion transitioning into subdiffusion in magnetically confined plasmas [24], also in the case of energetic particle trajectories traced in MHD simulation results featuring SMFRs separated by small-scale current sheets in the solar coronal plasma [17].

2. Motivation

le Roux and Zank [25] showed how one can derive from first principles a pitch-angle dependent fractional diffusion-advection kinetic equation in conservation form to model the anomalous diffusion of energetic particles interacting with SMFRs in the inner heliospheric supermagnetosonic solar wind on the basis of the standard focused transport equation. This was achieved by adjusting the approach of Sanchez et al. [26], applied to fluid turbulence in fusion plasmas, to this problem. The resultant anomalous transport theory of le Roux and Zank [25] suffers from the following shortcomings: (1) The asymptotic power law of a Lévy distribution was specified in the derivation to model the non-Gaussian statistics of the disturbed energetic particle trajectories in ordinary, momentum, and pitch-angle space generated when energetic particles interact with numerous SMFRs. The problem is that the second moment (variance) and higher moments of the Lévy distribution are infinite, indicating over-efficient non-local transport that is scale-free. (2) The theory does not naturally allow for a transition of anomalous transport to normal diffusion, or to a different anomalous diffusion state. An outline of a derivation is presented here in which an exponentially truncated Lévy distribution was specified instead, resulting in a tempered fractional diffusion-advection kinetic equation that addresses these two concerns.

3. Derivation of a Tempered Fractional Kinetic Transport Theory for Energetic Particle Interaction with SMFRs in the Large-Scale Solar Wind

3.1. Key Aspects of Derivation

To derive from first principles a kinetic transport theory to model the anomalous diffusion of energetic particles interacting with numerous coherent SMFR structures, including a transition to a normal diffusion or different anomalous diffusion state, requires a number of key elements:

(1) Since SMFRs contain intermittent strong perpendicular magnetic field component increments, as discussed above, strong disturbances are induced in, e.g., the perpendicular guiding center velocity ($\delta \boldsymbol{V}_g^I \approx v\mu\delta \boldsymbol{B}_I/B_0$, where v is the particle speed, μ is the cosine of the particle's pitch angle, and $\delta \boldsymbol{B}_I$ is the twist/island component of the SMFR magnetic field in the 2D plane perpendicular to the

guide/background magnetic field $\boldsymbol{B}_0$). Thus, the kinetic transport equation for anomalous transport is derived following a non-linear perturbation approach in which, e.g., the particle's disturbed perpendicular guiding center velocity correlation function decays, as a consequence of the decay in the correlation function of the twist component of the SMFR magnetic field ($\delta \boldsymbol{V}_g^I \propto \delta \boldsymbol{B}_I$), along the disturbed transverse guiding center trajectory (Lagrangian correlation function). In a traditional quasi-linear theory, the correlation function is specified to decay instead along the undisturbed guiding center trajectory (weak random wave turbulence). However, in the non-linear derivation, the Lagrangian correlation functions are more complicated. E.g., the disturbed perpendicular guiding center velocity correlation function also includes the variation in the perpendicular gradient of the ensemble-averaged particle distribution function along the disturbed transverse guiding center trajectory to form a correlation function with a triple product ensemble average. For comparison, in quasi-linear theory one finds the usual correlation function with a double product ensemble average only referring to the correlation of the disturbed perpendicular guiding center velocity. In [25], based on the approach of [26], it is shown how this complication can be overcome by switching to a Eulerian approach to deal with triple product ensemble averages. This approach is also followed in this paper.

(2) Non-Gaussian power-law statistics must be specified to model, e.g., the statistics of disturbances in the perpendicular guiding center velocity in response to the non-Gaussian power-law statistics associated with perpendicular magnetic field component increments of numerous SMFRs. For this purpose, the asymptotic of an exponentially truncated Lévy PDF is assumed for the rescaled disturbed transverse guiding center velocity component $\widetilde{\delta V_{g\iota}^I}$, e.g., which is given by

$$P\left(\widetilde{\delta V_{g\iota}^I}\right) = A_\alpha \frac{e^{-\lambda|\widetilde{\delta V_{g\iota}^I}|}}{|\widetilde{\delta V_{g\iota}^I}|^{1+\alpha}} \quad (0 < \alpha < 2), \tag{1}$$

where A_α is a constant that depends on the fractional index α and $1/\lambda$ is the characteristic transverse guiding center velocity above which the power-law PDF decays exponentially. The exponential rollover is the new element in the anomalous transport theory that was not included in the previous version of the theory presented in [25]. This new addition ensures that unphysically large disturbances in the perpendicular guiding center velocity, e.g., are avoided, thus overcoming the problem of infinite second or higher moments of the standard Lévy distribution that results in scale-free non-local spatial transport in the previous version of the theory [25]. It is this exponential truncation that enables a transition of, e.g., the anomalous perpendicular diffusion of energetic particle transport to more normal diffusion as the power-law statistics of the disturbed transverse guiding center velocities gets replaced by more Gaussian-like statistics beyond the critical scale $1/\lambda$.

This statistical description is further elaborated by introducing a simple scattering model of energetic particles by SMFR structures. When particles propagate inside coherent SMFRs structures, it is assumed that they have disturbed but coherent guiding center trajectories. Furthermore, it is assumed that particle scattering occurs when they propagate through the abrupt SMFR boundaries characterized by strong discontinuities in the perpendicular magnetic component increments, because of small-scale current sheets separating neighbouring SMFRs [13]. Accordingly, one can think of the particle waiting time (scattering time) Δt as the trapping time in or escape time from SMFR structures, and the perpendicular step size Δx_i as determined by the SMFR transverse crossing distance which is approximately equal to the perpendicular size of SMFRs. By defining the rescaled average disturbed transverse guiding center velocity inside a SMFR as

$$\widetilde{\delta V_{g\iota}^I} = \frac{\Delta x_i}{\Delta t^H}, \tag{2}$$

where the so-called Hurst index, $H = \beta/\alpha$, is the ratio of the fractional time index of the tempered fractional time derivative over the fractional spatial index of the tempered fractional perpendicular spatial derivative, a link is established between the guiding center velocity and the particle waiting time and step size. In the derivation this connection results in a separation of the original exponentially truncated Lévy PDF of equation (1) into an exponentially truncated Lévy PDF for the step size and an exponentially truncated PDF for the waiting time that can be considered to be independent:

$$P(\Delta x_i) = A_\alpha \frac{e^{-\lambda_x |\Delta x_i|}}{|\Delta x_i|^{1+\alpha}} \quad (0 < \alpha < 2),$$
$$P(\Delta t) = A_{\alpha H} \frac{e^{-\lambda_t (\Delta t)}}{(\Delta t)^{1+(2-\alpha H)}} \quad (0 < \alpha H = \beta < 2). \tag{3}$$

In this aspect the theoretical development is similar to the classical continuous-time random walk model for particle scattering in which the step-size and waiting-time PDFs are assumed to be independent [27]. Note that both PDFs include an exponential truncation so that beyond a critical step size and waiting time Gaussian-like statistics for the particle step size and Poisson-type statistics for the particle waiting time are gradually restored.

(3) As a consequence of the strong energetic particle response to SMFRs featuring intermittently strong transverse magnetic field component increments at SMFR boundaries, the spatial scale L over which the average energetic particle distribution $\langle f \rangle$ varies is such that $L \approx l_c$ where l_c is the spatial scale over which the disturbed transverse guiding center velocity correlation function decays along the disturbed guiding center trajectory. This means that the perpendicular gradient in the particle distribution about a correlation length removed from position x_i is such that

$$\frac{\partial \langle f \rangle}{\partial x_i}(x_i - \Delta x_i \approx x_i - l_c, t) \not\approx \frac{\partial \langle f \rangle}{\partial x_i}(x_i, t). \tag{4}$$

Thus, spatial lags in $\langle f \rangle$, and more generally, lags in time, momentum, and pitch angle, must be retained in this non-linear approach and cannot be ignored as is done in quasi-linear theory. Retaining these lags is a key element that enables modelling of anomalous diffusion of energetic particles in terms of fractional derivatives. Consequently, in the derivation of the new tempered anomalous transport equation, the exponentially truncated Lévy distributions result in transport terms for anomalous diffusion featuring tempered fractional derivatives containing non-local spatial, momentum, pitch-angle, and memory effects of particle transport that are taken into account by integration. For example, for $1 < \alpha < 2$ there appears in the perpendicular anomalous diffusion term a left Caputo tempered fractional derivative

$$\frac{{}^{C}\partial^{\alpha-1,\lambda_x} \langle f \rangle}{\partial x_i^{\alpha-1}}(x, t) = \frac{1}{\Gamma(2-\alpha)} \int_0^{x_i - x_{i0} > 0} d(\Delta x_i) \frac{e^{-\lambda_x (\Delta x_i)}}{(\Delta x_i)^{\alpha-1}} \frac{\partial \langle f \rangle}{\partial x_i}(x_i - \Delta x_i, t), \tag{5}$$

in which non-local transport effects are taken into account by integrating the perpendicular gradient of $\langle f \rangle$ over the interval $x_i - x_{i0}$. Notice also that the exponential rollover in the exponentially truncated step-size Lévy distribution (equation (3)) features in the tempered fractional derivative, thus ensuring that non-local transport effects beyond a certain critical spatial scale λ_x^{-1} is ignored. As a consequence, anomalous perpendicular diffusion on intermediate spatial scales undergoes a transition to more normal diffusion or a different anomalous diffusion state on large scale spatial scales [21, 24].

3.2. A Brief Outline of the Derivation

The derivation of the transport equation for tempered anomalous transport started with the standard focused transport equation for the evolution of the nearly gyrotropic energetic particle distribution $f(x_i, t, p, \mu)$ as a function of particle position x_i, time *t*, momentum *p* (speed *v*), and cosine of pitch angle μ given by

$$\frac{\partial f}{\partial t} + \langle V_{gi} \rangle_\phi \frac{\partial f}{\partial x_i} + \langle \frac{dp}{dt} \rangle_\phi \frac{\partial f}{\partial p} + \langle \frac{d\mu}{dt} \rangle_\phi \frac{\partial f}{\partial \mu} = 0, \tag{6}$$

where the particle's gyro-phase-averaged guiding center velocity is expressed as

$$\langle V_{gi} \rangle_\phi = U_i + v\mu b_i, \tag{7}$$

the gyro-phase-averaged momentum rate of change is given by

$$\frac{1}{p}\langle \frac{dp}{dt} \rangle_\phi = -\frac{1}{2}(1-\mu^2)\frac{\partial U_i}{\partial x_i} - \frac{1}{2}(3\mu^2-1)b_i b_j \frac{\partial U_i}{\partial x_j} + \mu\left(\frac{qE_i}{p} - \frac{1}{v}\frac{dU_i}{dt}\right) b_i, \tag{8}$$

and the gyro-phase-averaged μ-rate of change is

$$\langle \frac{d\mu}{dt} \rangle_\phi = \frac{1}{2}(1-\mu^2)\left[\mu \frac{\partial U_i}{\partial x_i} - 3\mu b_i b_j \frac{\partial U_i}{\partial x_j} + 2\left(\frac{qE_i}{p} - \frac{1}{v}\frac{dU_i}{dt}\right) b_i + v\frac{\partial b_i}{\partial x_i}\right]. \tag{9}$$

In equations (7)-(9), U_i refers to the non-uniform solar wind flow velocity, μ is the cosine of the particle's pitch angle, b_i indicates the unit vector along the non-uniform heliospheric magnetic field, E_i is the electric field, and dU_i/dt is the acceleration of the solar wind flow velocity as seen following the flow. Equations (8) and (9) indicate that energetic particles change energy and direction because of non-uniformities in the solar wind flow. This includes the divergence of the solar wind flow (first term in (8) and (9)), the parallel component of the solar wind shear flow (second term in (8) and (9)), and the parallel component of the non-inertial force associated with the acceleration of the solar wind (fourth term in (8) and (9)). Furthermore, energy and direction changes also occur due to the parallel component of the electric field force in the solar wind (third term in (8) and (9)), and direction changes are produced by the divergence of the magnetic field unit vector in the focusing/mirroring term (last term in (9)).

A quasi-linear theory type perturbation analysis was used to distinguish the response of the energetic particle distribution to the large-scale solar wind flow and magnetic field on long time scales, and to the intermediate scale, random flow and magnetic fields of individual SMFR structures on intermediate time scales. An important distinction between this perturbation approach for non-propagating coherent SMFR structures and the standard perturbation approach for random propagating MHD wave turbulence is that, while the large-scale average of the SMFR flow velocity over numerous SMFRs is specified to be zero, this was not assumed to be necessarily the case for the large-scale average of gradients in the SMFR velocity listed in equations (8) and (9). This extension facilitates the modeling of energetic particle acceleration during a situation where multiple SMFRs are simultaneously contracting, such as expected during early-time magnetic reconnection at a large-scale current sheet. Furthermore, $E_i b_i$ in equations (8) and (9) refers to the parallel electric field (the component of the electric field that is non-motional or non-ideal). However, on intermediate scales (MHD scales) the motional electric field is typically considered to be the dominant electric field component and one should think of the motional electric field components parallel and perpendicular to the average/guide magnetic field in contracting and merging SMFRs to be responsible for accelerating energetic particles [4]. In this theory, the motional electric field in individual SMFRs, namely $-\delta \boldsymbol{U}_I \times \delta \boldsymbol{B}_I$ parallel to the average/guide field $\boldsymbol{B}_0$, is used to model the effect of the parallel SMFR electric field on energetic particles on intermediate spatial and time scales [25] because both the SMFR flow velocity $\delta \boldsymbol{U}_I$ and the twist component of the magnetic field $\delta \boldsymbol{B}_I$ are modelled to exist predominantly in the 2D plane perpendicular to $\boldsymbol{B}_0$ (quasi-2D model for dynamic SMFRs).

Implementation of the perturbation analysis method, and the above-mentioned three key elements needed for the development of a tempered anomalous transport theory, led to the derivation of a pitch-angle dependent tempered fractional diffusion-advection equation in conservation form for modeling the response of energetic particles to numerous dynamic SMFRs (for more detail, see [25]).

3.3. The New Tempered Fractional Kinetic Equation for Anomalous Transport

Compactly expressed, the equation for the evolution of the average energetic particle distribution function $\langle f\rangle(x_i,t,p,\mu)$ is

$$\begin{aligned}
&\frac{\partial\langle f\rangle}{\partial t}+\frac{\partial}{\partial x_i}\left(\left[\langle V_{gi}\rangle_{SW}+\langle V_{gi}\rangle_I\right]\langle f\rangle\right)+\frac{1}{p^2}\frac{\partial}{\partial p}\left(p^2\left[\langle\frac{dp}{dt}\rangle_{SW}+\langle\frac{dp}{dt}\rangle_I\right]\langle f\rangle\right)\\
&\qquad+\frac{\partial}{\partial\mu}\left(\left[\langle\frac{d\mu}{dt}\rangle_{SW}+\langle\frac{d\mu}{dt}\rangle_I\right]\langle f\rangle\right)\\
&=\frac{\partial^{1-\beta,\lambda_t}}{\partial t^{1-\beta}}\frac{\partial}{\partial x_{i\perp}}\left(\kappa_\perp^{I[\beta,\alpha]}\frac{\partial^{\alpha-1,\lambda_x}\langle f\rangle}{\partial^{\alpha-1}x_i}\right)+\frac{\partial^{1-\beta,\lambda_t}}{\partial t^{1-\beta}}\frac{1}{p^2}\frac{\partial}{\partial p}\left(p^2\kappa_p^{I[\beta,\alpha]}\frac{\partial^{\alpha-1,\lambda_p}\langle f\rangle}{\partial^{\alpha-1}p}\right)\\
&\qquad+\frac{\partial^{1-\beta,\lambda_t}}{\partial t^{1-\beta}}\frac{\partial}{\partial\mu}\left(\kappa_\mu^{I[\beta,\alpha]}\frac{\partial^{\alpha-1,\lambda_\mu}\langle f\rangle}{\partial^{\alpha-1}\mu}\right). \qquad (10)
\end{aligned}$$

Equation (10) is valid for fractional indices in the range $0<\beta=\alpha H<2$, and $0<\alpha<2$. On the left-hand-side of equation (10) are the terms containing particle fluxes in ordinary, momentum and in μ-space, respectively. One can see the decomposition of the fluxes into that generated by particle interaction with the background solar wind flow and magnetic field (terms with the subscript "SW") and the fluxes produced by particle interaction with the mean SMFR flow, magnetic field, and motional electric field (terms with subscript "I"). The expressions for the advection velocities were derived from the basic expressions in equations (7) - (9) (for more information, see [25]). These transport terms have normal derivatives and are unchanged from those presented in [25].

What is new are the anomalous diffusion terms on the right-hand-side that include tempered fractional time, spatial, momentum, and μ-derivatives (or integrals) whereas previously in [25] they contained normal fractional derivatives/integrals. With this extension of the anomalous transport theory, one can model naturally a transition from anomalous diffusion to more normal diffusion at later times (if the tempered fractional time derivatives/integrals are neglected), or a transition to a different anomalous diffusion state, as discussed above. It can be shown that when the fractional indices $\beta\to 1$, and $\alpha\to 2$, the tempered fractional time derivatives/integrals vanish by becoming one, while the tempered fractional spatial, momentum, and μ-derivatives/integrals become normal so that normal diffusion terms are recovered. As an example, we provide more details of the tempered fractional momentum derivative that can be expressed as

$$\frac{\partial^{\alpha-1,\lambda_p}\langle f\rangle}{\partial^{\alpha-1}p}=\left(-{}_pD_{p_0}^{\alpha-1,\lambda_p}+{}_{p_0}D_p^{\alpha-1,\lambda_p}\right)\langle f\rangle, \qquad (11)$$

where for $1<\alpha<2$, the left Riemann-Liouville tempered fractional derivative ${}_{p_0}D_p^{\alpha-1,\lambda_p}$ is equivalent to the left Caputo tempered fractional derivative ${}_{p_0}^{C}D_p^{\alpha-1,\lambda_p}$ for $\langle f\rangle(p=p_0)=0$ so that

$$\begin{aligned}
{}_{p_0}D_p^{\alpha-1,\lambda_p}\langle f\rangle(x_i,t,p,\mu) &= {}_{p_0}^{C}D_p^{\alpha-1,\lambda_p}\langle f\rangle(x_i,t,p,\mu)\\
&=\frac{1}{\Gamma(2-\alpha)}\int_0^{p-p_0>0}d(\Delta p)\frac{e^{-\lambda_p(\Delta p)}}{(\Delta p)^{\alpha-1}}\frac{\partial\langle f\rangle}{\partial p}(x_i,t,p-\Delta p,\mu), \qquad (12)
\end{aligned}$$

and in the same way the right Riemann-Liouville tempered fractional derivative ${}_pD_{p_0}^{\alpha-1,\lambda_p}$ is equal to the right Caputo tempered fractional derivative ${}_p^{C}D_{p_0}^{\alpha-1,\lambda_p}$ whereby

$$\begin{aligned}
{}_pD_{p_0}^{\alpha-1,\lambda_p}\langle f\rangle(x_i,t,p,\mu) &= {}_p^{C}D_{p_0}^{\alpha-1,\lambda_p}\langle f\rangle(x_i,t,p,\mu)\\
&=\frac{1}{\Gamma(2-\alpha)}\int_{p-p_0<0}^{0}d(\Delta p)\frac{e^{-\lambda_p|\Delta p|}}{|\Delta p|^{\alpha-1}}\frac{\partial\langle f\rangle}{\partial(-p)}(x_i,t,p-\Delta p,\mu). \qquad (13)
\end{aligned}$$

For $0 < \alpha < 1$, the left Riemann-Liouville tempered fractional derivative ${}_{p_0}D_p^{\alpha-1,\lambda_p}$ is equal to the left Riemann-Liouville tempered integral ${}_{p_0}I_p^{1-\alpha,\lambda_p}$ if $\langle f\rangle(p=p_0)=0$ so that

$$
\begin{aligned}
{}_{p_0}D_p^{\alpha-1,\lambda_p}\langle f\rangle(x_i,t,p,\mu) &= {}_{p_0}I_p^{1-\alpha,\lambda_p}\langle f\rangle(x_i,t,p,\mu)\\
&= \frac{1}{\Gamma(1-\alpha)}\int_0^{p-p_0>0} d(\Delta p)\frac{e^{-\lambda_p(\Delta p)}}{(\Delta p)^{\alpha}}\langle f\rangle(x_i,t,p-\Delta p,\mu),
\end{aligned} \tag{14}
$$

and in the same way the right Riemann-Liouville tempered fractional derivative ${}_pD_{p_0}^{\alpha-1,\lambda_p}$ is equal to the right Riemann-Liouville fractional integral ${}_pI_{p_0}^{1-\alpha,\lambda_p}$ given by

$$
\begin{aligned}
{}_pD_{p_0}^{\alpha-1,\lambda_p}\langle f\rangle(x_i,t,p,\mu) &= {}_pI_{p_0}^{1-\alpha,\lambda_p}\langle f\rangle(x_i,t,p,\mu)\\
&= \frac{1}{\Gamma(1-\alpha)}\int_{p-p_0<0}^{0} d(\Delta p)\frac{e^{-\lambda_p|\Delta p|}}{|\Delta p|^{\alpha}}\langle f\rangle(x_i,t,p-\Delta p,\mu).
\end{aligned} \tag{15}
$$

Furthermore, the tempered fractional time derivatives in (10) refers to a left Riemann-Liouville tempered fractional derivative ${}_0D_t^{1-\beta,\lambda_t}$. For $1<\beta=\alpha H<2$, ${}_0D_t^{1-\beta,\lambda_t}$ is equivalent to a left Riemann-Liouville tempered fractional integral ${}_0I_t^{\beta-1,\lambda_t}$, that is

$$
\begin{aligned}
\frac{\partial^{1-\beta,\lambda_t}}{\partial t^{1-\beta}}\langle f\rangle(x_i,t,p,\mu) &= {}_0D_t^{1-\beta,\lambda_t}\langle f\rangle(x_i,t,p,\mu) = {}_0I_t^{\beta-1,\lambda_t}\langle f\rangle(x_i,t,p,\mu)\\
&= \frac{1}{\Gamma(\beta-1)}\int_0^{t} d(\Delta t)\frac{e^{-\lambda_t(\Delta t)}}{(\Delta t)^{2-\beta}}\langle f\rangle(x_i,t-\Delta t,p,\mu),
\end{aligned} \tag{16}
$$

whereas for $1<\beta=\alpha H<1$, ${}_0D_t^{1-\beta,\lambda_t}$ is equivalent to the left Caputo tempered fractional derivative ${}_0^CD_t^{1-\beta,\lambda_t}$ if $\langle f\rangle(t=0)=0$ so that

$$
\begin{aligned}
{}_0D_t^{1-\beta,\lambda_t}\langle f\rangle(x_i,t,p,\mu) &= {}_0^CD_t^{1-\beta,\lambda_t}\langle f\rangle(x_i,t,p,\mu)\\
&= \frac{1}{\Gamma(\beta)}\int_0^{t} d(\Delta t)\frac{e^{-\lambda_t(\Delta t)}}{(\Delta t)^{1-\beta}}\frac{\partial\langle f\rangle}{\partial t}(x_i,t-\Delta t,p,\mu).
\end{aligned} \tag{17}
$$

The transport coefficients in the anomalous diffusion terms can be compactly expressed as follows:

$$
\begin{aligned}
\kappa_p^{I[\beta,\alpha]} &= C(\beta,\alpha)\sigma_p^{\alpha} = C(H,\alpha)\left\langle\left|\delta\left(\frac{dp}{dt}\right)_I\right|^{\alpha}\right\rangle\langle\tau_{cor}\rangle^{\alpha-\beta},\\
\kappa_\mu^{I[\beta,\alpha]} &= C(\beta,\alpha)\sigma_\mu^{\alpha} = C(H,\alpha)\left\langle\left|\delta\left(\frac{d\mu}{dt}\right)_I\right|^{\alpha}\right\rangle\langle\tau_{cor}\rangle^{\alpha-\beta},\\
\kappa_\perp^{I[\beta,\alpha]} &= C(\beta,\alpha)\sigma_{x_i}^{\alpha} = (H,\alpha)\langle|\delta V_{gi}^{I}|^{\alpha}\rangle\langle\tau_{cor}\rangle^{\alpha-\beta},
\end{aligned} \tag{18}
$$

where $C(\beta,\alpha)=h(\beta,\alpha)(1-\alpha)\Gamma(\beta)/(2\cos(\alpha\pi/2))$, and $\langle\tau_{cor}\rangle$ is the average correlation time (the mean time for the disturbed particle guiding-center velocity correlation function to decay along the guiding-center trajectory through SMFR structures), modeled as the average time to cross SMFR structures (the assumption is that particles scatter when crossing SMFR boundaries, thus causing the decay in the correlation function as discussed above). In [25] estimates for the correlation time can be found. In the anomalous diffusion coefficient expressions of equation (18), $\delta(dp/dt)_I$, e.g., is the deviation in the particle momentum rate of change in individual SMFRs from the mean momentum rate of change $\langle dp/dt\rangle_I$ determined from interaction with many SMFRs. The expression for $\langle|\delta(dp/dt)|^{\alpha}\rangle$, was derived from the basic expression given by equation (8) (see [25]) under the limitation that $|\ \delta(dp/dt)_I| \gg |\langle dp/dt\rangle_I|$, implying that the anomalous momentum diffusion term in (10) are stronger than the advection term in momentum space during particle interaction with SMFRs (term with subscript "I" on the left-hand-side in (10)). This implies the assumption that particle energization by the variance in the SMFR flow compression, e.g., is more important than energization by the average SMFR flow compression. This approach was also followed to derive expressions for the other anomalous diffusion coefficients in (18). Note that normal diffusion coefficients can be recovered by letting $\beta \rightarrow 1$, and $\alpha \rightarrow 2$ in the anomalous diffusion coefficient expressions in equation (18), whereby $C(\beta,\alpha) \rightarrow 1/2$ because $h(\beta,\alpha) \rightarrow 1$. Then, e.g.,

$$\kappa_p^{I[\beta=1,\alpha=2]} = \sigma_p^2 = \langle\ \delta(dp/dt)_I^2\ \rangle\langle\tau_{cor}\rangle = \langle(\Delta p)^2/\Delta t\rangle\,, \tag{19}$$

where $\sigma_p^2$is the square of the standard deviation of the PDF for $\delta(dp/dt)_I$.

4. Summary

Evidence is mounting that SMFRs as part of a non-propagating quasi-2D MHD turbulence component have a strong presence in the inner heliosphere as has been predicted for many years. The PDF of the magnetic field component increments of these nonlinear structures is characterized by non-Gaussian power-law statistics with strong tails generated by abrupt SMFR boundaries separated by secondary small-scale current sheets. Thus, there is good reason to think that the spatial, momentum, and pitch-angle transport of energetic particles propagating through solar wind regions containing numerous SMFRs is anomalous on intermediate time scales when particles follow field lines. The idea of intermediate time-scale anomalous transport is supported by evidence of particle trapping in SMFRs in the perpendicular direction accompanied by coherent momentum gain, and evidence of spatial and momentum superdiffusion of particles propagating in a field of SMFRs separated by small-scale current sheets. It is expected that at later times there should be a transition to normal diffusive behavior or a different anomalous diffusion state.

le Roux and Zank [25] showed previously how one can derive from first principles a pitch-angle dependent fractional diffusion-advection kinetic equation in conservation form to model the anomalous diffusion of energetic particles interacting with SMFRs in the inner heliospheric supermagnetosonic solar wind on the basis of the standard focused transport equation. This equation suffers from following shortcomings: (1) The asymptotic power law of a Lévy distribution was specified in the derivation to model the non-Gaussian statistics of the disturbed energetic particle trajectories in ordinary, momentum, and pitch-angle space generated when energetic particles interact with numerous SMFRs. The problem is that the second moment (variance) and higher moments of the Lévy distribution are infinite, indicating over-efficient non-local transport that is scale-free. (2) The theory does not naturally allow for a transition of anomalous transport to normal diffusion, or to a different anomalous diffusion state.

These shortcomings have been addressed in this paper by specifying instead the asymptotic of an exponentially truncated Lévy distribution to model the statistics of the disturbed particle trajectories. This led to the derivation of a new pitch-angle dependent tempered fractional diffusion-advection equation in conservation form, presented in brief outline here, in which the advection terms remain normal (i.e., feature normal derivatives) and unchanged from the advection terms presented in [25]. What is new are the anomalous diffusion terms that include tempered fractional time, spatial, momentum, and μ-derivatives (or integrals) whereas previously in [25] they featured normal fractional derivatives/integrals. The tempered derivatives/integrals exhibit the same exponential rollover present in the exponentially truncated Lévy distributions. With this extension of the anomalous transport theory one can model naturally a transition from anomalous diffusion to more normal diffusion (if the tempered fractional time derivatives are neglected), or a transition to a different anomalous diffusion state.

A more detailed presentation of the derivation of the tempered factional transport equation will be presented in a near future publication, including presenting the derivation of a tempered fractional Parker transport equation which holds the promise for finding simplified analytical solutions that can be used to evaluate more sophisticated numerical solutions in solar wind situations where near isotropic pitch-angle distributions for energetic particles can be assumed.

Acknowledgements
J. A. le Roux acknowledges support from NASA Grants 80NSSC21K1319, 80NSSC20K1783, 80NSSC19K0276, and from NSF Grant EPSCoR RII-Track-1 Cooperative Agreement OIA-1655280.

References
[1] Zank G P and Matthaeus W H 1992 *J. Plasma Phys.* **48** 85
[2] Zank G P and Matthaeus W H 1993 *Physics Fluids A: Fluid Dynamics* **5** 257
[3] Zank G P, Adhikari L, Hunana P, Shiota D, Bruno R and Telloni D 2017 *Astrophys. J.* **835** 147
[4] Dmitruk P, Matthaeus W H and Seenu N 2004 *Astrophys. J.* **617** 667
[5] Matthaeus W H, Goldstein M L and Roberts D A 1990 *J. Geophys. Res.* **95** 20673
[6] Bieber J W, Wanner W and Matthaeus W H 1996 *J. Geophys. Res.* **101** 2511
[7] Hu Q, Zheng J, Chen Y, le Roux J A and Zhao L 2018 *Astrophys. J. Suppl. Ser.* **239** 12
[8] Zhao L-L, Zank G P, Adhikari L, Nakanotani M, Telloni D and Carbone F 2020 *Astrophys. J.* **898** 113
[9] Pecora F et al. *Month. Not. Royal Astron. Soc.* **508** 2114
[10] Ruffolo D, Matthaeus W H and Chuychai P 2003 *Astrophys. J.* **597** L169
[11] Trenchi L, Bruno R, Telloni D, D'amicis R, Marcucci M F, Zurbuchen T H and Weberg M 2013, *Astrophys. J.* **770** 11
[12] Drake J F, Swisdak M and Fermo R 2013 *Astrophys. J.* **763** L5
[13] Greco A, Matthaeus W H, Servidio S, Chuychai P and Dmitruk P 2009 *Astrophys. J.* **691** L111
[14] Zheng J and Hu Q 2018 *Astrophys. J.* **852** L23
[15] Isliker H, PisokasT, Vlahos L and Anastasiadis A 2017 *Astrophys. J.* **849**, 35
[16] Isliker H, Vlahos L and Constantinescu D 2017 *Phys. Rev. Lett.* **119** 045101
[17] Isliker H, Archontis V and Vlahos L 2019 *Astrophys. J.* **882**, 57
[18] Nakanotani M, Zank G P and Zhao L-L 2022 *Front. Astron. Space Sci.* **9** 954040
[19] Sandev T, Chechkin A and Metzler R 2018 *Conf. Proc. CFPM (Ohrid, Macedonia)* (Ohrid:Soc. Physicists Macedonia) p 10
[20] Ito K and Miyazaki S 2003 *Prog. Theor. Phys.* **110** 875
[21] Vallaeys V, Tyson R C, Lane W D, Deleersnijder E and Hanert E 2017 *J. R. Soc. Interface* **14** 20160889
[22] Fisk L A and Axford W I 1969 *Sol. Phys.* **7** 486
[23] Qin G, Matthaeus W H and Bieber J W 2002 Astrophys. Lett. **578** L117
[24] Cartea A and Del-Castillo-Negrete D 2007 *Phys. Rev. E* **76** 041105
[25] le Roux J A and Zank G P 2021 *Astrophys. J.* **913** 84
[26] Sanchez R, Carreras B A, Newman D E, Lynch V E and van Milligan B 2006 *Phys. Rev.* **74** 016305
[27] Montroll E W and Weiss J 1965 *J. Math. Phys.* **6** 167

Journal of Physics: Conference Series **2544** (2023) 012006 doi:10.1088/1742-6596/2544/1/012006

Understanding the heating mechanism of the solar active region atmosphere in chromosphere

Mehmet Sarp Yalim[1], Gary Zank[1,2] Christian Beck[3], Debi Prasad Choudhary[4], Avijeet Prasad[5], Qiang Hu[1,2], and Makayla Frisse[1]

[1] Center for Space Plasma and Aeronomic Research, The University of Alabama in Huntsville, 320 Sparkman Drive, Huntsville, AL 35805, USA
[2] Department of Space Science, The University of Alabama in Huntsville, 320 Sparkman Drive, Huntsville, AL 35805, USA
[3] National Solar Observatory (NSO), 3665 Discovery Drive, Boulder, CO 80303, USA
[4] Department of Physics and Astronomy, California State University, Northridge (CSUN), Northridge, CA 91330-8268, USA
[5] Rosseland Centre for Solar Physics, University of Oslo, Postboks 1029, Blindern NO-0315, Oslo, Norway

E-mail: msy0002@uah.edu,garyp.zank@gmail.com,cbeck@nso.edu, debiprasad.choudhary@csun.edu,avijeet.prasad@astro.uio.no,qh0001@uah.edu, mpf0004@uah.edu

Abstract. Understanding the mechanisms underlying the heating of the solar atmosphere is a fundamental problem in solar physics. In this paper, we present an overview of our research on understanding the heating mechanism of the solar active region atmosphere in chromosphere. We investigate Joule heating due to the dissipation of currents perpendicular to the magnetic field by the Cowling resistivity using a data-constrained analysis based on observational and tabulated theoretical/semi-empirical solar atmosphere model data. As target region, we focus on a sunspot umbral light bridge where we find that this heating mechanism plays an important role and is also highly dynamic.

1. Introduction

One of the fundamental problems in solar physics is to determine the physical processes that maintain the thermal structure of the solar atmosphere [1]. There are various mechanisms attributed to the energy transfer throughout the solar atmosphere. In this paper, we investigate a mechanism that heats the solar active region atmosphere in chromosphere: Joule heating due to the dissipation of electric currents by magnetic resistivity, namely Cowling resistivity.

The energy transfer in the chromosphere can be attributed to a number of mechanisms such as Alfvén waves [2, 3], nanoflares [4, 5], Ellerman bombs (EBs) [6, 7], spicules [8, 9], and magneto-acoustic shocks [10]. [11, 12] suggested that the heating of the chromosphere is due to the dissipation of acoustic waves, which has, however, been questioned by [13, 14, 15].

Another candidate to heat the solar chromosphere would be resistive Ohmic dissipation [16, 17]. The plasma in the chromosphere is not fully ionized and ions and neutrals exist simultaneously. Cowling resistivity follows the interactions between ions and neutrals. It is a function of the plasma bulk density, temperature, magnetic field, as well as the ion and electron number densities. The Joule heating mechanism that we focus on results from the dissipation of electric currents that are perpendicular to the magnetic field due to Cowling resistivity. This mechanism is represented by the generalized Ohm's law as a function of Cowling resistivity and electric current perpendicular to the magnetic field. We perform a data-constrained analysis [18, 19] to calculate the Cowling resistivity and hence the Joule heating rate in a solar active region atmosphere based on tabulated data of stratified bulk plasma, ion and electron number densities, and temperature profiles from the tabulated data of five different theoretical/semi-empirical solar atmosphere models, namely the Maltby M model [20] for sunspot umbrae, the VAL C model [21] and the Harvard-Smithsonian Reference Atmosphere (HSRA) model [22] for quiet Sun, the VAL F model [21] for bright filament networks, and the Ding & Fang model [23] for sunspot penumbrae, in combination with the application of a non-force-free-field (NFFF) magnetic field extrapolation technique [24, 25, 26] to the photospheric vector magnetic field data observed by the Helioseismic and Magnetic Imager (HMI) [27] magnetogram onboard Solar Dynamics Observatory *(SDO)* [28] to obtain the magnetic field. In addition, we also use temperature data from the inversion of spectroscopic data obtained by the Interferometric BI-dimensional Spectrometer (IBIS) instrument [29] at the ground-based Dunn Solar Telescope *(DST)* to further constrain our data analysis, and hence decrease the reliance of our analysis on the tabulated data from static solar atmosphere models.

The occurrence of dynamic phenomena in the chromosphere and transition region has been attributed to plasma heating by the formation of current sheets when a discontinuity in the 3D magnetic field arises [30, 31]. We focus on discontinuities in the magnetic field topology which produce electric currents that can heat the plasma in those regions by dissipation through the Cowling resistivity. Such currents can occur at all locations with strong gradients in the magnetic field strength or orientation. These conditions are prevalent for instance near light bridges (LBs) inside the umbra of sunspots, for magnetic flux emergence into a field-free or magnetic environment, near polarity inversion lines or near magnetic reconnection sites like EBs. In this paper, we analyze Joule heating in a sunspot umbral LB in NOAA AR 12002 on 2014 March 13 from 20:44 to 21:00 UT by applying the data-constrained analysis mentioned above.

Section 2 presents an overview of our data-constrained analysis to investigate the effects of Cowling resistivity on heating in the weakly-ionized chromosphere. Section 3 shows the results obtained for the Joule heating of a sunspot umbral LB. Finally, section 4 presents our conclusions.

2. Data-constrained analysis of Joule heating in the chromosphere

To describe the interaction of the chromospheric plasma with the magnetic field, and its dependence on the degree of collisional coupling, we apply a quasi-MHD single fluid theory complemented with a generalized Ohm's law as in [32]. Here, we will present a brief overview about how to calculate the Cowling resistivity. The details of the chromosphere model utilized can be found in [18].

Our main focus is on the calculation and effects of Cowling resistivity. The anisotropic dissipation of currents due to the presence of Cowling resistivity can be seen in the induction and energy equations given in Eq. 1 and Eq. 2, respectively, that illustrate the impact of the presence of neutrals through the terms that contain η_C:

$$\frac{\partial \mathbf{B}}{\partial t} + \nabla \cdot (\mathbf{vB} - \mathbf{Bv}) + \nabla \times \eta J_{\parallel} + \nabla \times \eta_C J_{\perp} = 0, \quad (1)$$

and

$$\frac{\partial E}{\partial t} + \nabla \cdot \Big[(E + p + \frac{B^2}{8\pi})\mathbf{v} - \frac{\mathbf{B}}{4\pi}(\mathbf{v} \cdot \mathbf{B})\Big] - \nabla \cdot (\mathbf{B} \times \eta J_{\parallel}) - \nabla \cdot (\mathbf{B} \times \eta_C J_{\perp}) = \rho(\mathbf{v} \cdot \mathbf{g}) + S_{NA}. \quad (2)$$

Here, η is the Coulomb resistivity, η_C is the Cowling resistivity, $J_{\parallel}$ and $J_{\perp}$ are the components of current density parallel and perpendicular to the magnetic field $\mathbf{B}$, E, p, ρ, $\mathbf{v}$, and $\mathbf{g}$ are specific total energy, thermal pressure, density, velocity, and gravitational acceleration, respectively, and S_{NA} is the combination of non-adiabatic source terms corresponding to viscous heating, shock heating, thermal conduction, radiative transfer, and coronal heating.

To calculate the Cowling resistivity, η_C, we solve the following relation between the Cowling and Coulomb resistivities:

$$\frac{\xi_n^2 B_0^2}{\alpha_n} = \eta_C - \eta, \tag{3}$$

where B_0 is the magnetic field strength, ξ_n is an estimate for the neutral fraction which is a function of density and temperature, and α_n is also a function of density and temperature.

To calculate the Coulomb and Cowling resistivities, we need the magnetic field $\mathbf{B}$, the plasma bulk density ρ and temperature T as well as the ion and electron number densities, n_i and n_e, in the chromosphere. We calculate the 3D spatial variation of T in the vicinity of the target structure of interest (i.e., in this paper, a sunpot umbral LB) in the chromosphere by using the Calcium Inversion based on a Spectral ARchive (CAISAR) code [33, 34, 35] that inverts the spectroscopic data obtained from *DST*/IBIS. For the 1D variations of bulk plasma density ρ, electron number density n_e and the total hydrogen number density n_H with height above the photosphere, we use tabulated data from the Maltby-M, VAL C, VAL F, Ding & Fang, and HSRA atmosphere models. We then calculate n_i from n_e and n_H. Finally, the magnetic field $\mathbf{B}$ is taken from the NFFF magnetic field extrapolation results.

It should be noted that we also have the option to use either the 1D stratified temperature profile with height from the tabulated model data or temperature data from inversion results of observations to test their effects on the analysis results.

After calculating the Cowling resistivity, we then calculate the electric currents from the gradients of the 3D extrapolated magnetic field topology to calculate the Joule heating from the generalized Ohm's law as $\eta_C J_\perp^2$.

3. Analysis of Joule heating in a sunspot umbral light bridge

As a primary target structure in our analysis, we focus on a sunspot umbral LB. LBs are elongated bright structures that are usually present during the early stages of sunspot formation or the late stages of sunspot decay and can have an umbral, penumbral or granular morphology [36]. They are relatively long-lived (hours to days) and less dynamic structures which makes them particularly suitable for heating by our proposed Joule heating mechanism. In [19], we demonstrated that the primary heating mechanism in a sunspot LB caused by flux emergence is Joule heating due to dissipation of currents by the Cowling resistivity based on our data-constrained analysis that involves data from observations and atmosphere models.

In this paper, we build on our analysis in [19], which utilized only magnetic field from observations while the rest of the plasma variables were taken from the tabulated data given by the Maltby M model including temperature, and present the effects of using the 3D spatial variation of temperature based on observational data on the Joule heating of the same LB.

We analyze observations of the leading sunspot in NOAA AR 12002 on 2014 March 13 from 20:44 to 21:00 UT, combining data from vector magnetograms observed by *SDO*/HMI SHARP data series and temperature data obtained by the inversion of spectroscopic data measured by *DST*/IBIS using the CAISAR inversion code. For more details about the data specifics, we refer the interested reader to [19].

Figure 1 (left panel) shows the field of view (FOV) of the HMI SHARP magnetogram (outer box) as well as the IBIS FOV (solid box) and a smaller FOV (dashed box) focusing on the LB. Our results are based on this smaller FOV.

In order to replace the tabulated temperature data with the observational temperature data in our analysis, it is necessary to assure the consistency between the observational temperature data and the rest of the plasma variables taken from tabulated solar atmosphere data. We have

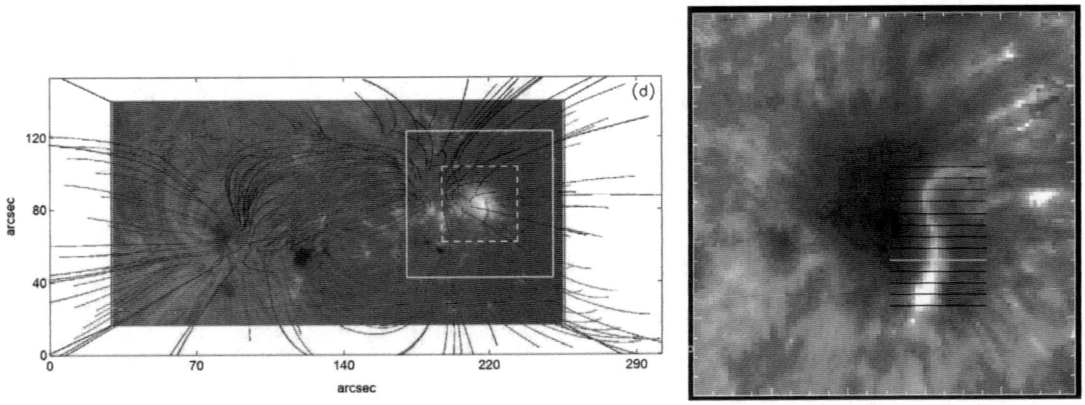

Figure 1. (Left panel) Field lines derived from the non-force-free field (NFFF) extrapolation overlaid on a composite image of the vertical component of the magnetic field and the AIA 171 Å image for the SHARP field of view (FOV) of AR 12002. The solid and dashed white squares correspond to the IBIS FOV and the smaller FOV shown in the right panel, respectively [19]; (Right panel) Spatial distribution of the LTE temperature from *DST*/IBIS data at $z = 0.36$ Mm height above the photosphere in AR 12002. The white horizontal line marks slit No. 5 that crosses the LB.

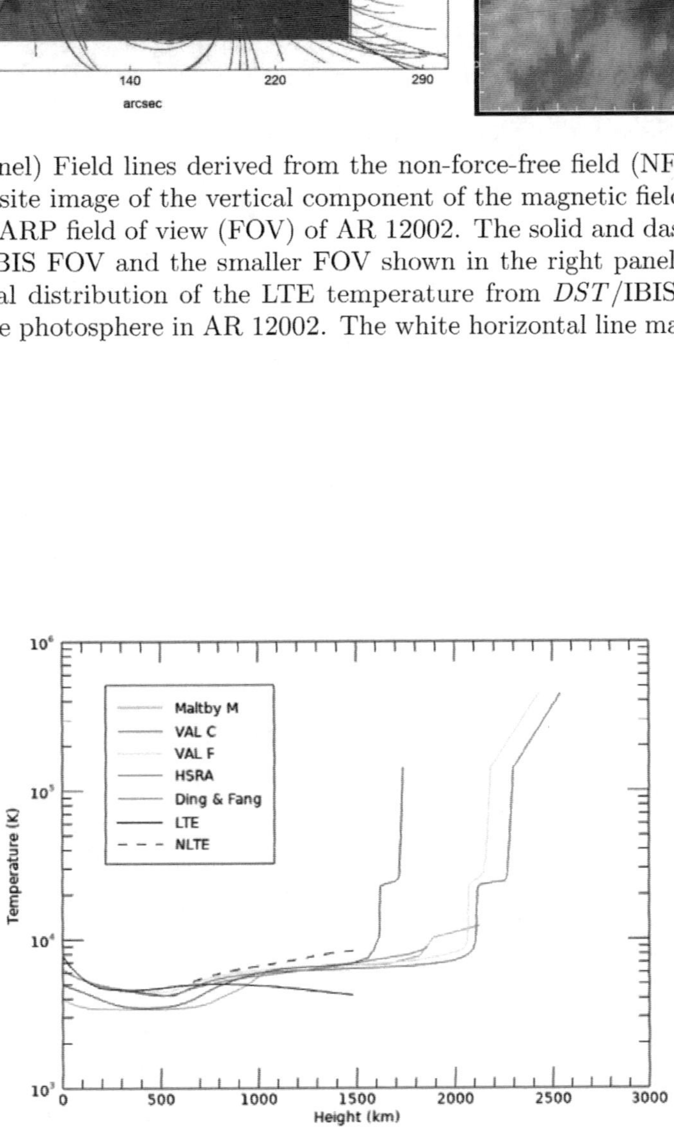

Figure 2. Temperature profiles in the chromosphere vs height above the photosphere from tabulated data of five solar atmosphere models. The LTE and NLTE temperature profiles are obtained from an inversion of spectroscopic data from *DST*/IBIS at the location where the height-averaged Joule heating rate distribution is maximum along the white slit shown in Figure 1 (right panel).

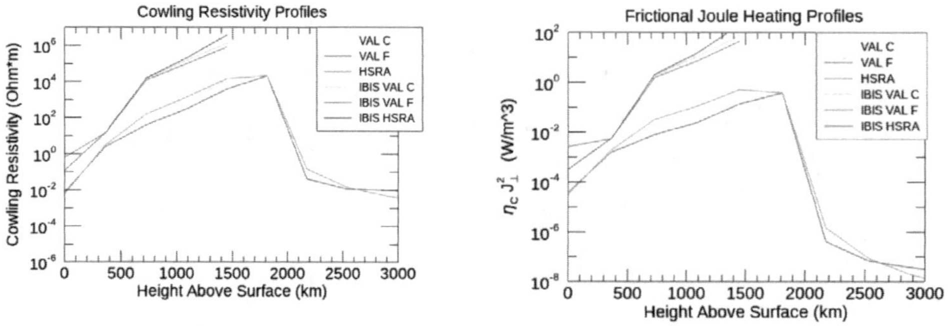

Figure 3. Variations of the maximum values of (Left panel) η_C and (Right panel) Joule heating rate profiles with height above the photosphere, calculated from VAL C, VAL F, and HSRA models with temperature from model or IBIS data. Both HMI and IBIS data correspond to AR 12002 at 21:00 UT.

five solar atmosphere models that model different types of regions of the solar active region atmosphere. In order to determine the model to use in our analysis, we compare the IBIS temperature data with the 1D stratified temperature variation with height from all five models as shown in Figure 2. We make this comparison at a representative location on the LB which is the location where the height-averaged Joule heating rate distribution is maximum along the white slit, which is shown in Figure 1 (right panel), that crosses approximately the center of the LB.

While we used Maltby-M model in our previous analysis in [19], the temperature data from IBIS that is obtained from the non-local thermodynamic equilibrium (NLTE) version of the inversion model, which is more accurate than the LTE version of the inversion model, agrees with the VAL F model the most, and also with the VAL C and HSRA models within 0-1,485 km height above the photosphere. We note that 1,485 km is the maximum height at which we obtain IBIS temperature data from the CAISAR inversion code.

Let us first have a look at the variations of Cowling resistivity and the associated Joule heating in the dashed box in Figure 1. Figure 3 shows the variations of the maximum values of η_C and the Joule heating rate with height above the photosphere using tabulated data or IBIS data for temperature in our analysis. For all three model selections to calculate the Cowling resistivity, namely VAL F, VAL C, and HSRA, using the 3D NLTE temperature distribution from IBIS instead of the 1D tabulated temperature profile leads to a Joule heating rate which is 2-3 orders of magnitude higher. This result already shows the impact of using observational temperature data in our analysis. We should, however, note that the locations of the maximum values of η_C and the Joule heating rate are not necessarily located on the LB at each height.

Let us now focus specifically on the LB. Figure 4 (left panel) shows the horizontal variations in the LTE and NLTE temperature data, averaged in height, across the center of the LB at the white slit. The middle and right panels in the same figure show the horizontal variations of the current perpendicular to the magnetic field $J_\perp$, η_C, and the Joule heating rate along the white slit. While η_C is calculated by plasma values from the VAL F model in both panels, tabulated temperature values are used in the middle panel whereas the inverted temperature values from IBIS are utilized in the right panel. While the distributions of all three quantities are qualitatively similar in both middle and right panels, quantitatively the maximum value of the Joule heating rate is 7 times higher when the IBIS temperature is used. In addition, the location of the peak of the NLTE temperature from IBIS coincides with the peak of the Joule

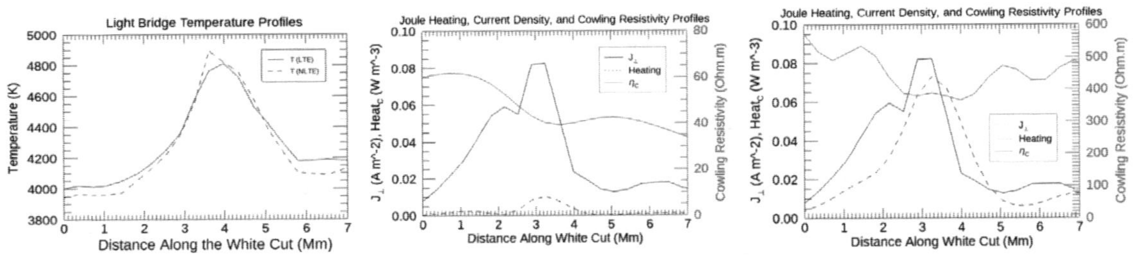

Figure 4. (Left panel) Horizontal variations in temperature from IBIS data, averaged in height, along the white slit shown in Figure 1 (right panel) across the LB; (Middle & right panels) Joule heating rate, $J_\perp$, and η_C profiles along the white slit. All three quantities are averaged in height up to 1,485 km. $J_\perp$ has been scaled up by a factor of three and Joule heating rate has been scaled up by a factor of two. η_C is calculated based on the VAL F model using (Middle panel) tabulated temperature and (Right panel) observational temperature (NLTE) from IBIS. Both HMI and IBIS data correspond to AR 12002 at 20:48 UT.

heating rate which supports the role of our proposed Joule heating mechanism in the heating of the LB. If we look into the variations of the two terms that Joule heating is a function of, namely η_C and $J_\perp$, we see that η_C makes a dip and $J_\perp$ has a peak at the peak location of the NLTE temperature from IBIS. This result indicates that the variation of $J_\perp$ plays a major role in our proposed heating mechanism.

The results shown in Figure 4 are based on height averaging of the quantities across the white slit. *How do η_C and the associated Joule (Ohmic) heating vary with height in the vicinity of the LB?* Figure 5 presents the variations of these quantities at three different locations in the vicinity of the LB. The highly dynamic nature of the Joule heating can be clearly seen as the profiles in these locations look very different from each other both qualitatively and quantitatively.

Finally, let us make a quantitative comparison of the total Joule heating rate calculated by integrating the individual Joule heating rate values at each height at the location where the NLTE temperature from IBIS makes a peak along the white slit. Table 1 presents the Joule heating rate values obtained by using tabulated plasma variable values from each atmospheric model at two different times, namely 20:48 and 21:00 UT. Accordingly, the Joule heating rates are consistently larger at 21:00 UT than their values at 20:48 UT for each model which demonstrates that there is an ongoing heating event in the LB based on our proposed heating mechanism. The last two rows of the table show the results based on using tabulated and IBIS temperature data, respectively in our analysis. The total Joule heating rate at this location is 147 times larger at both times when we utilize the temperature variation from IBIS in our calculations. This result shows the importance of using observational data for temperature instead of tabulated data in our analysis and its impact on the results of our analysis.

4. Conclusions

In this paper, we investigate a mechanism that heats the solar active region atmosphere: Joule heating due to the dissipation of currents perpendicular to the magnetic field lines by Cowling resistivity in the chromosphere.

We give an overview of a data-constrained analysis to calculate the Cowling resistivity and the associated Joule heating. Our analysis can be constrained by both observational data (in particular, magnetic field from NFFF extrapolations applied to *SDO*/HMI magnetograms and temperature from the inversion of spectroscopic data from *DST*/IBIS) and tabulated data from five distinct theoretical/semi-empirical solar atmosphere models corresponding to different

Journal of Physics: Conference Series **2544** (2023) 012006 doi:10.1088/1742-6596/2544/1/012006

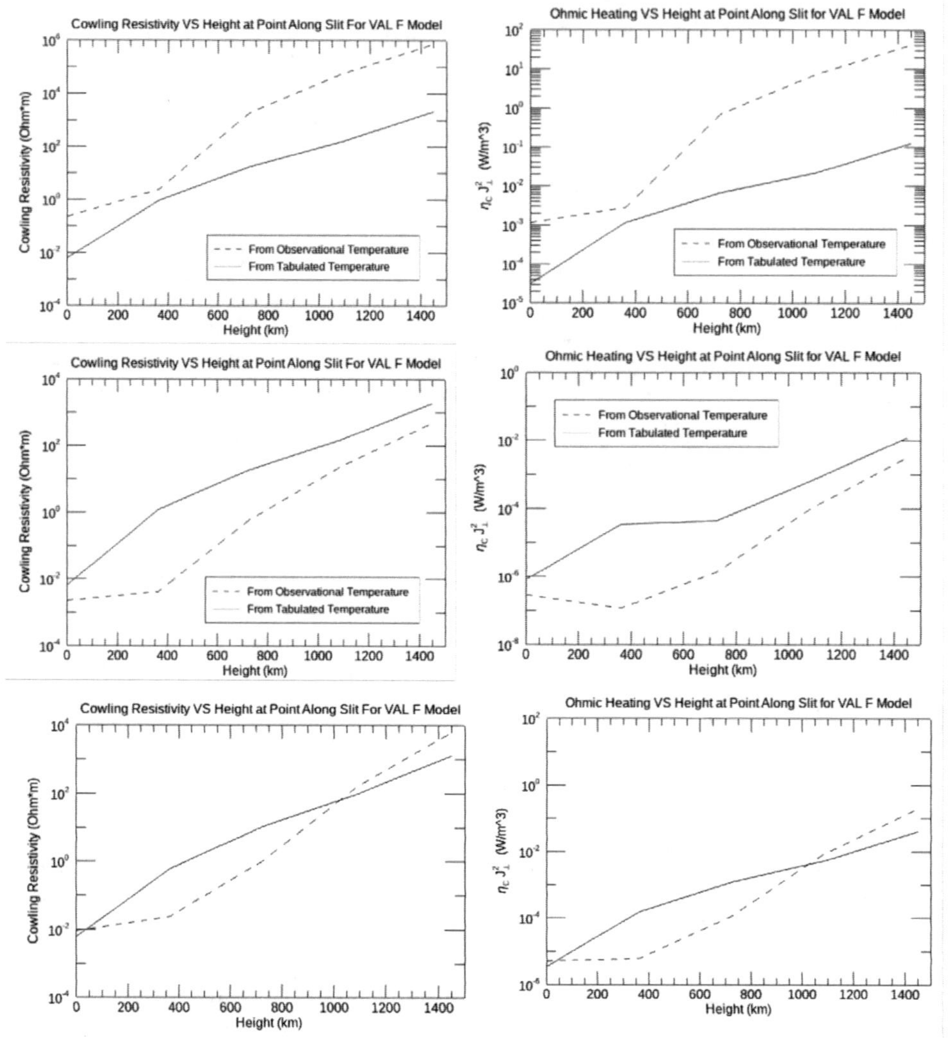

Figure 5. (Top panel) (Left) η_C and (Right) Joule heating rate variations with height at the location in the dashed line box in Figure 1 (left panel) where the NLTE temperature is maximum between the heights 1,000-1,500 km above the photosphere; (Middle panel) (Left) η_C and (Right) Joule heating rate variations with height at the center of the white slit in Figure 1 (right panel); (Bottom panel) (Left) η_C and (Right) Joule heating rate variations with height in the vicinity of the LB on the southernmost black slit in Figure 1 (right panel).

regions of the active region atmosphere. We analyze Joule heating over a sunspot umbral LB by using observational data for temperature and tabulated data for the rest of the plasma variables that are needed to calculate the Cowling resistivity from the VAL F model. VAL F is selected for our analysis since its 1D stratified temperature profile with height agrees the most with the NLTE temperature profile calculated from IBIS data at a representative point on the LB. We see a consistent increase of up to three orders of magnitude in the maximum values of Cowling resistivity and Joule heating rate at each height that are calculated using observational temperature instead of tabulated temperature from the VAL F, VAL C, and HSRA models. We observe an alignment of the height-averaged LTE and NLTE temperature, $J_\perp$, and the Joule

Model	**Height range (km)**	**Joule heating rate (W/m^3) (20:48 UT)**	**Joule heating rate (W/m^3) (21:00 UT)**
Maltby M	0-2,500 km	1.2894	1.6812
VAL C	0-2,500 km	1.3542	1.8088
VAL F	0-2,500 km	0.3635	0.4847
HSRA	0-2,500 km	0.6846	0.9073
Ding & Fang	0-2,500 km	3.106	4.0749
VAL F	0-1,485 km	0.111	0.146
VAL F + IBIS	0-1,485 km	16.378	21.345

Table 1. The total Joule heating rate values between 0-2,500 km or 0-1,485 km height at the maximum height-averaged Joule heating location on the LB across the white cut in Figure 1 (right panel) for each model at 20:48 and 21:00 UT.

heating rate peaks across the LB. This demonstrates a strong dependence of heating on current, because the Joule heating rate makes a peak despite η_C having a local minimum at that location. We also observe higher values of up to 7 times for height-averaged Cowling resistivity and Joule heating rate when calculated with observational temperature data compared to those values when calculated with tabulated temperature from the VAL F model. We see that the total Joule heating rates at 21:00 UT are consistently greater than their values at 20:48 UT, and thus we can conclude that the LB experiences an ongoing heating event due to this mechanism. In particular, we observe that the total Joule heating rate values found using the NLTE temperature data from IBIS with the VAL F model are two orders of magnitude greater than their counterparts found using the VAL F model including the temperature. Finally, we find that the Joule heating rate is highly dynamic both on and in the vicinity of the LB from its variations with height at different locations. All the above observations support the importance of constraining our analysis using observational data. In particular, constraining temperature using observational temperature data from IBIS in our analysis is a major step forward from our last study [19]. In future work, we plan to calculate temperature and internal energy enhancements within the LB with respect to the umbral surroundings for quantitative validation for the same LB, and repeat our analysis on different target structures within the chromosphere.

Acknowledgments

We acknowledge support from the NSF EPSCoR RII-Track-1 Cooperative Agreements OIA-1655280 and OIA-2148653. We also acknowledge support from the NSF awards AGS-2020703 and AGS-2230633. M.F. acknowledges support from the NSF REU program under award number AGS-1950831. Any opinions, findings, and conclusions or recommendations expressed in this material are those of the author(s) and do not necessarily reflect the views of the National Science Foundation.

References

[1] Narain U and Ulmschneider P 1996 *Space Sci. Rev.* **75** 453
[2] van Ballegooijen A A, Asgari-Targhi M, Cranmer S R and DeLuca E E 2011 *Astrophys. J.* **736** 3
[3] Sakaue T and Shibata K 2020 *Astrophys. J.* **900** 120
[4] Priest E R, Chitta L P and Syntelis P 2018 *Astrophys. J. Lett.* **862** L24
[5] Syntelis P and Priest E R 2020 *Astrophys. J.* **891** 52
[6] Rutten R J 2016 *A & A* **590** A124

[7] Chen Y, Tian H, Peter H, Samanta T, Yurchyshyn V, Wang H, Cao W, Wang L and He J 2019 *Astrophys. J. Lett.* **875** L30
[8] De Pontieu B, McIntosh S W, Hansteen V H and Schrijver C J 2009 *Astrophys. J.* **701** L1
[9] Beck C, Rezaei R, Puschmann K G and Fabbian D 2016 *Sol. Phys.* **291** 2281
[10] De Pontieu B, McIntosh S, Martinez-Sykora J, Peter H and Pereira T M D 2015 *Astrophys. J. Lett.* **799** L12
[11] Ulmschneider R, Schmitz F, Kalkofen W and Bohn H U 1978 *A & A* **70** 487
[12] Kalkofen W 2007 *Astrophys. J.* **671** 2154
[13] Athay R G and Holzer T E 1982 *Astrophys. J.* **255** 743
[14] Beck C, Khomenko E, Rezaei R and Collados M 2009 *A & A* **507** 453
[15] Beck C, Rezaei R and Puschmann K G 2012 *A & A* **544** A46
[16] Parker E N 1983 *Astrophys. J.* **264** 635
[17] Tritschler A, Uitenbroek H and Reardon K 2008 *Astrophys. J.* **686** L45
[18] Yalim M S, Prasad A, Pogorelov N V, Zank G P and Hu Q *Astrophys. J. Lett.* **899** L4
[19] Louis R E, Prasad A, Beck C, Choudhary D P and Yalim M S 2021 *A & A* **652** L4
[20] Maltby P, Avrett E H, Carlsson M, Kjeldseth-Moe O, Kurucz R L and Loeser R 1986 *Astrophys. J.* **306** 284
[21] Vernazza J E, Avrett E H and Loeser R 1981 *Astrophys. J. Suppl. Ser.* **45** 635
[22] Gingerich O, Noyes R W, Kalkofen W and Cuny Y 1971 *Sol. Phys.* **18** 347
[23] Ding M D and Fang C 1989 *A & A* **225** 204
[24] Hu Q and Dasgupta B 2008 *Sol. Phys.* **247** 87
[25] Hu Q, Dasgupta B, Choudhary D P and Büchner J 2008 *Astrophys. J.* **679** 848
[26] Hu Q, Dasgupta B, DeRosa M L, Büchner J, and Gary G A 2010 *J. Atmos. Sol. Terr. Phys.* **72** 219
[27] Schou J *et al* 2012 *Sol. Phys.* **275** 229
[28] Pesnell W D, Thompson B J and Chamberlin P C 2012 *Sol. Phys.* **275** 3
[29] Cavallini F 2006 *Sol. Phys.* **236** 415
[30] Solanki S K, Lagg A, Woch J, Krupp N and Collados M 2003 *Nature* **425** 692
[31] Bahauddin S M, Bradshaw S J and Winebarger A R 2021 *Nat. Astron.* **5** 237
[32] Leake J E and Arber T D 2006 *A & A* **450** 805
[33] Beck C, Rezaei R and Puschmann K G 2013 *A & A* **549** A24
[34] Beck C, Choudhary D P, Rezaei R and Louis R E 2015 *Astrophys. J.* **798** 100
[35] Beck C, Gosain S and Kiessner C 2019 *Astrophys. J.* **878** 60
[36] Louis R E, Beck C and Choudhary D P 2020 *Astrophys. J.* **905** 153

Slow solar wind modeling of the Metis/Solar Orbiter – Parker Solar Probe quadrature

L Adhikari[1], G P Zank[1], D Telloni[2], L -L Zhao[1], and A Pitna[3]
[1]Center for Space Plasma and Aeronomic Research (CSPAR) and Department of Space Science, University of Alabama in Huntsville, Huntsville, AL 35899, USA
[2]National Institute for Astrophysics—Astrophysical Observatory of Torino Via Osservatorio 20, I-10025 Pino Torinese, Italy
[3]Charles University, Faculty of Mathematics and Physics, V Holešovičkách 2, 180 00 Prague 8, Czech Republic

E-mail: la0004@uah.edu

Abstract. In January 2021, Metis/SolO and PSP formed a quadrature from which the slow solar wind was able to be measured from the extended solar corona (3.5 – 6.3 $R_\odot$) to the very inner heliosphere (23.2 $R_\odot$). Metis/SolO remotely measured the coronal solar wind, finding a speed of 96 – 201 kms^{-1}, and PSP measured the solar wind in situ, finding a speed of 219.34 kms^{-1}. Similarly, the normalized cross-helicity and the normalized residual energy measured by PSP are 0.96 and -0.07. In this manuscript, we study the evolution of the proton entropy and the turbulence cascade rate of the outward Elsässer energy during this quadrature. We also study the relationship between solar wind speed, density and temperature, and their relationship with the turbulence energy, the turbulence cascade rate, and the solar wind proton entropy. We compare the theoretical results with the observed results measured by Metis/SolO and PSP.

1. Introduction

Since the launch of the Parker Solar Probe (PSP) and the Solar Orbiter (SolO) in 2018 and 2020 respectively, these two spacecraft have been measuring the solar wind plasma between the Sun and the Earth. During encounter 8, PSP entered a region of sub-Alfvénic solar wind for the first time [1], marking a major accomplishment of the PSP mission [2], and provided important information about turbulence near the Alfvén surface [1,3]. Similarly, the Metis coronagraph onboard the SolO [4] remotely measures the solar wind plasma in the extended solar corona [e.g., 5-7]. The dissipation of low-frequency turbulence is regarded as a promising mechanism for heating the solar corona and accelerating the solar wind [8-10, 18]. PSP and SolO provide insights on these interesting problems from an observational perspective. In addition, the PSP magnetometer [11] and plasma [12], and the SolO magnetometer [13] and plasma [14]

Journal of Physics: Conference Series **2544** (2023) 012007 doi:10.1088/1742-6596/2544/1/012007

data sets are also used to study the radial evolution of turbulence in the inner heliosphere [15,16,17].

Telloni et al [18] investigated the first alignment between PSP and SolO, when PSP was at 0.1 au at 4:00 UT on 2020-9-27, and SolO was at 1 au at 22:19 UT on 2020-10-01. Assuming ballistic propagation, Telloni et al were able to make in situ measurements of the same plasma parcel with a 1.5 hour long interval traveling from 0.1 au to 1 au. Similarly, Telloni et al [19] identified a PSP – SolO quadrature, in which the PSP during E7 was located at 0.11 au and at 18:59 UT on 2021-01-18 entered the plane of sky (POS) measured by Metis coronagraph onboard SolO on 2021-01-17 at 16:30 UT. In this case, Metis/SolO and PSP measured the same plasma parcel remotely and in situ (indicated as interval #1 in [19]) traveling from the extended solar corona (3.3 – 6.3 $R_\odot$) to the very inner heliosphere (23.2 $R_\odot$).

Using potential field source surface (PFSS) extrapolation [20], Telloni et al [19] found that the source region of the PSP interval #1 plasma is the equatorial extension of the southern polar coronal hole. The average solar wind speed of this interval plasma is about 219.34 kms^{-1}, and the observed normalized cross-helicity is about 0.96, indicating that this is Alfvénic slow solar wind [6,19]. Similarly, the angle between the mean magnetic field and the mean solar wind speed is measured to be $\theta_{UB} \sim 165°$, which indicates that the observed turbulence corresponding to this interval is slab turbulence.

Adhikari et al [6] used the solar wind (SW) + NI MHD turbulence transport model equations [10] to compare the theoretical model and the joint PSP – SolO observations for the first time, from the extended solar corona to the very inner heliosphere. Adhikari et al compared the theoretical results of the solar wind speed and the solar wind density with the measured values derived from Metis/SolO and PSP, and compared the theoretical results of 2D and slab turbulence energy and the corresponding correlation lengths with the measured results derived PSP. Similarly, Biondo et al [21] modeled the PSP – SolO quadrature using the Reverse in situ and MHD Approach (RIMAP), which is a hybrid analytical-numerical method performing data-driven reconstructions of the Parker spiral. Biondo et al found good agreement between the numerical, and the Metis/SolO and PSP measured speed and density.

In this manuscript, we do not discuss the SW + NI MHD turbulence transport model equations since the model equations are presented in Adhikari et al [6]. Using the model results of Adhikari et al [6], we calculate the solar wind proton entropy and the turbulent heating rate of the outward Elsässer energy, and then consider their relationship with solar wind parameters (solar wind speed, solar wind density, and proton temperature) and the turbulence energy. Similarly, we investigate the relationship between solar wind parameters. We discuss our results in Section 2, and present a discussion and conclusions in Section 3.

2. Results

Adhikari et al [6] solved the SW + NI MHD turbulence transport model equations from 3.3 $R_\odot$ to 30 $R_\odot$ using the boundary conditions shown in Table 1. In the table, $\langle z^{\infty\pm2}\rangle$ denotes the 2D outward/inward Elsässer energy and $L_\infty^\pm$ represents the corresponding energy weighted correlation length. E_D^∞ is the 2D residual energy and L_D^∞ is the energy weighted correlation length of the residual energy. $\langle z^{*+2}\rangle$ is the slab outward Elsässer energy, and L_*^+ is the energy weighted correlation length of $\langle z^{*+2}\rangle$. U is the solar wind speed, n is the proton density, and T is the proton temperature.

2D	Values	Slab	Values	SW	Values
$\langle z^{\infty\pm2}\rangle$ (km^{2}s^{-2})	10^5	$\langle z^{*+2}\rangle$ (km^{2}s^{-2})	6×10^3	U (kms^{-1})	45.13
E_D^∞ (km^{2}s^{-2})	2×10^3	L_*^+ (km^{3}s^{-2})	9.1×10^7	n (cm^{-3})	8×10^5
$L_\infty^\pm$ (km^{3}s^{-2})	3.03×10^9	-	-	T (K)	7×10^5
L_D^∞ (km^{3}s^{-2})	1.6×10^8	-	-	-	-

Table 1. Boundary values at 3.3 R$_\odot$ for the turbulent quantities and the solar wind parameters. The same boundary condition is chosen for $\langle z^{\infty+2}\rangle$ and $\langle z^{\infty-2}\rangle$.

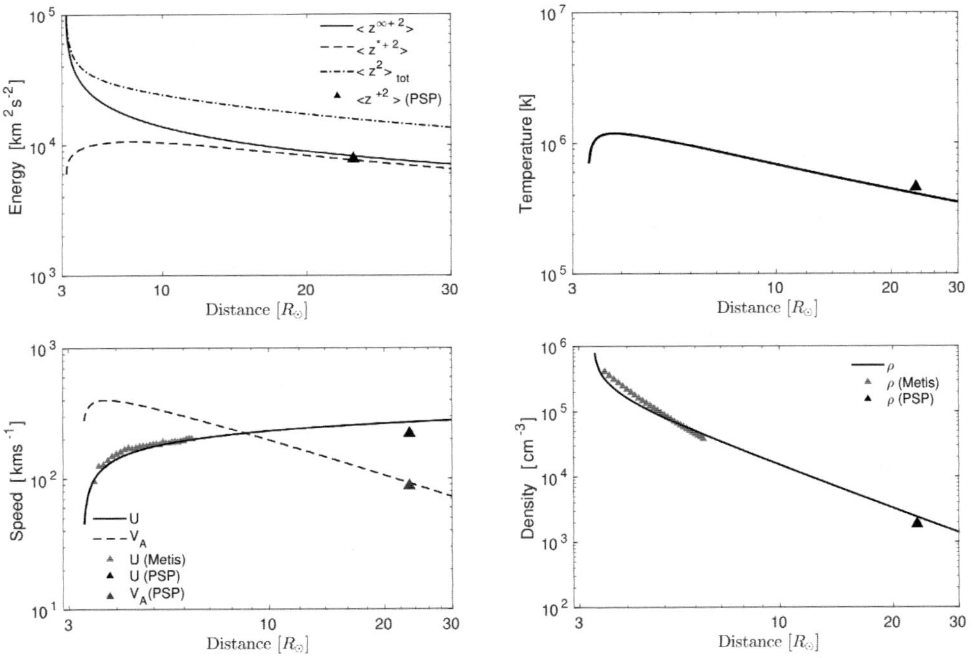

Figure 1. Comparison between the theoretical and observed outward Elsässer energy (top-left), proton temperature (top-right), solar wind speed and Alfvén velocity (bottom-left), and proton density (bottom-right) as a function of distance. The solid, dashed, and dashed-dotted-dashed curves represent the theoretical results. The red triangle denotes remotely measured speed and density by Metis/SolO (Telloni et al [19]), and the black and blue triangles denote the PSP measured values during encounter 7 (E7) (Reproduced from Adhikari et al [6]).

The top-left panel of Figure 1 displays the theoretical 2D (solid curve), slab (dashed), and 2D+slab (dashed-dotted-dashed curve), and the observed transverse (black triangle) outward Elsässer energy as a function of distance. Obviously, the theoretical $\langle z^{\infty+2}\rangle$ in the extended solar corona is larger than the theoretical $\langle z^{*+2}\rangle$, where $\langle z^{\infty+2}\rangle$ decreases with increasing distance [1], while $\langle z^{*+2}\rangle$ first increases and then decreases. The dissipated turbulence energy heats the solar corona, resulting in a coronal temperature of about 10^6 K, as shown in the top-right panel of Figure 1. The proton temperature increases abruptly from 7×10^5 K at 3.3 $R_\odot$ to 1.2×10^6 K at $\sim$ 3.75 $R_\odot$, mainly due to the dissipation of quasi-2D turbulence (see also Zank et al [9]). The hot solar corona produces a large pressure gradient that generates a large thermal force, which acts against the gravitational force, and drives the solar wind from a subsonic to a supersonic speed (bottom-left panel of Figure 1). The theoretical solar wind speed (solid curve) and the measured speed remotely by Metis/SolO (red triangles) increases rapidly between 3.3 $R_\odot$ and $\sim$ 5 $R_\odot$, and then the theoretical speed increases gradually, eventually being very close to the PSP measured solar wind speed (black triangle). The theoretical Alfvén velocity (dashed curve) increases from 281 kms^{-1} at 3.3 $R_\odot$ to $\sim$ 400 kms^{-1} at 3.75 $R_\odot$, and then decreases monotonically with increasing distance, also being very close to the PSP measured Alfvén velocity (blue triangle). In the bottom-right panel, the theoretical proton density (solid curve) is consistent with the remotely measured proton density of Metis/SolO (red triangle) and the in situ measured proton density of PSP (black triangle).

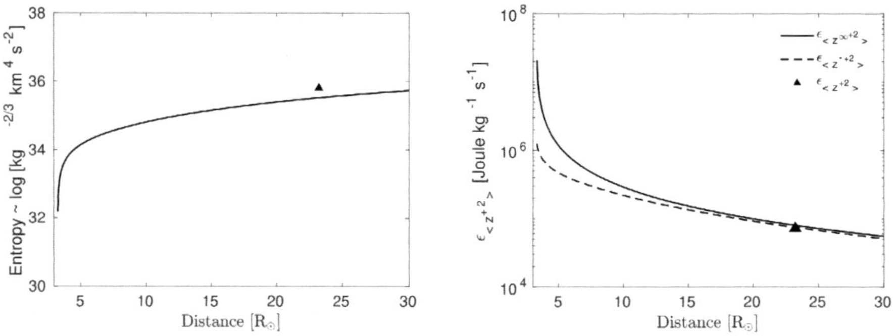

Figure 2. Left: Comparison between the theoretical (solid curve) and observed (black triangle) proton entropy as a function of heliocentric distance. Right: Comparison between the theoretical 2D (solid curve) and slab (dashed curve), and the observed (black triangle) heating rate of the outward Elsässer energy with increasing heliocentric distance.

Entropy is an important thermodynamic quantity, defined as $S = c_v \log(P/\rho^\gamma)$, where c_v is the specific heat capacity, P is the thermal pressure, ρ is the proton mass density, and $\gamma = 5/3$ is the polytropic index. Entropy in the solar wind plasma has been studied since the late 1980s (Whang et al [22,23]). Whang et al found that, as a result of shock heating, entropy increases with distance. Recently, Adhikari et al [24] put forward the

[1] The theoretical $\langle z^{\infty-2}\rangle$ exhibits the same radial profile as that of the $\langle z^{\infty+2}\rangle$, yielding the 2D normalized cross-helicity $\sigma_c^\infty = 0$ from 3.3 $R_\odot$ and 30 $R_\odot$ (see Adhikari et al [6]).

idea that the increase of entropy in the heliosphere is caused by turbulence dissipation. Here, we show the theoretical entropy from the extended solar corona to the very inner heliosphere, and compare with the observed entropy measured by PSP (left panel of Figure 2). Similar to the proton temperature, entropy increases rapidly within a small height from 3.3 $R_\odot$ to $\sim$ 4 $R_\odot$, which may be related to the rapid dissipation of 2D turbulence, and then increases gradually with distance.

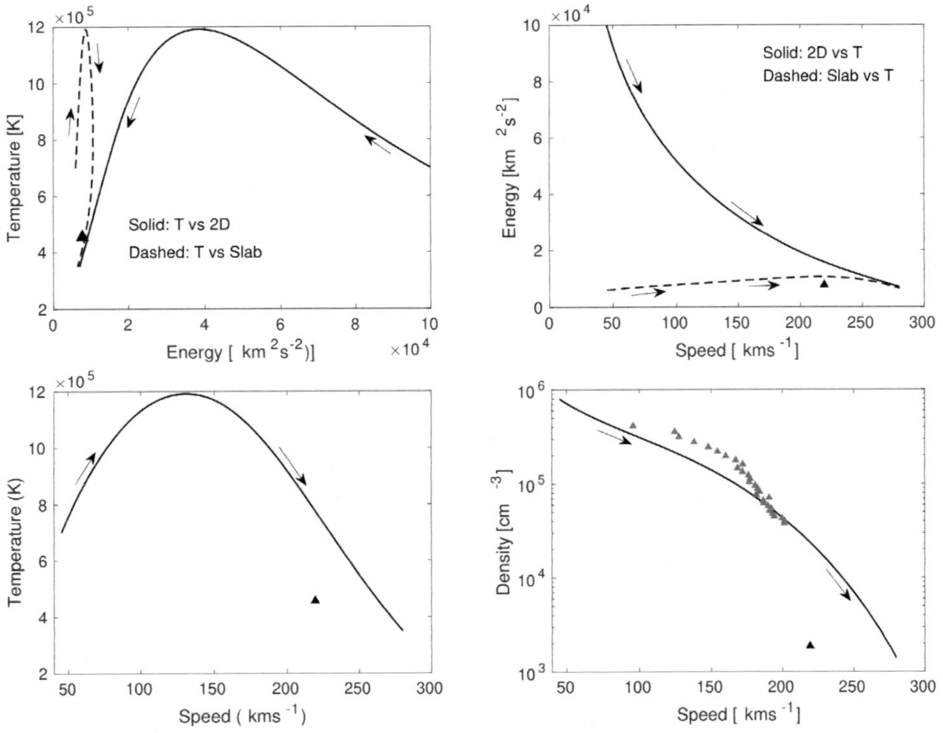

Figure 3. Top-left: Proton temperature as a function of outward Elsässer energy. Top-right: Outward Elsässer energy as a function of solar wind speed. Bottom-left: Proton temperature as a function of solar wind speed. Bottom-right: Proton density as a function of solar wind speed. The solid and dashed curves are the theoretical results. The black and red triangles are the observed results. The arrow denotes the direction from the start to end point.

A turbulence cascade simply means that the large-scale (kinetic+magnetic) energy is transferred into energy on smaller and smaller and at the very smallest scales is finally dissipated into heat energy. This is one of the well established ideas to explain the heating of the solar wind/corona. In the right panel of Figure 2, we plot the turbulent cascade rate/heating rate of the outward Elsässer energy as a function of distance. Here, the theoretical 2D (solid curve) and slab (dashed curve) turbulence cascade rates are calculated from the nonlinear dissipation terms of the 2D and slab turbulence (see [16], [25]). The observed cascade rate (black triangle) is calculated by

$\epsilon^+ = \langle z^{+2}\rangle^{3/2}/[C_K \log(1/k_{inj}\lambda^+)]^{3/2}\lambda^+$ (see [26]), where $\langle z^{+2}\rangle$ and λ^+ are the observed outward Elsässer energy and the corresponding correlation length, $k_{inj} = 1.07 \times 10^{-9}$ km^{-1} is the injection wavenumber [27], and C_K is the Kolmogorov constant. We use $C_K = 4.5$. Obviously, in the extended solar corona, the theoretical 2D cascade rate is larger than the theoretical slab cascade rate, supporting the idea that 2D turbulence is the main component responsible for heating the solar corona [9].

The top-left panel of Figure 3 shows the relationship between the proton temperature and the outward Elsässer energy. The solid curve denotes the theoretical T vs $\langle z^{\infty+2}\rangle$, the dashed curve denotes the theoretical T vs $\langle z^{*+2}\rangle$, and the black triangle denotes the observed result. The arrow indicates the direction of the curve from the start point to the end point. At first, $\langle z^{\infty+2}\rangle$ is inversely proportional to T until the maximum temperature (T_{max}), and then $\langle z^{\infty+2}\rangle$ is proportional to T. The theoretical T vs $\langle z^{*+2}\rangle$ curve shows that T and $\langle z^{*+2}\rangle$ are proportional until T_{max}, then they are inversely proportional, and then the two are directly proportional again. The theoretical T vs $\langle z^{*+2}\rangle$ curve approaches the observed result closely, in comparison to the theoretical T vs $\langle z^{\infty+2}\rangle$ curve. The top-right panel of Figure 3 shows that the theoretical $\langle z^{\infty+2}\rangle$ (solid curve) is inversely proportional to U, whereas, the theoretical $\langle z^{*+2}\rangle$ (dashed curve) is positively correlated with U until 215 kms^{-1}, close to the observed result, and then they are negatively correlated.

The bottom left panel of Figure 3 displays the relationship between the proton temperature and the solar wind speed. T and U are initially correlated until T_{max}, and then both are inversely correlated. Maksimovic et al [28] in their Figure 7, using the first orbit PSP and Helios 2 data sets, found that the proton temperature and the solar wind speed are proportional. However, in their Figures 4 and 5, they grouped the solar wind speed and the proton temperature according to the range of solar wind speeds, showing that the slow solar wind speed increases and the proton temperature decreases with distance, indicating that the proton temperature and the solar wind speed are anticorrelated, similar to our results. In a similar study, the anticorrelation between the electron temperature and the slow solar wind speed has also been observed by Halekas et al [29] using PSP measurements. In the bottom right panel of Figure 3, the theoretical (solid curve) and observed (red and black triangles) results show that the proton density decreases as a function of solar wind speed.

We plot the proton temperature as a function of proton entropy in the left panel of Figure 4. Initially, the proton temperature increases with proton entropy, and then decreases with increasing entropy. The right panel of Figure 4 shows that the proton density is anticorrelated with proton entropy, and the theoretical result is close to the observed result. This of course follows from the anti-correlated of ρ and U ($\rho U r^2=$ const).

The top-left panel of Figure 5 shows the relationship between the 2D/slab outward Elsässer energy and the corresponding heating rate. The theoretical $\langle z^{\infty+2}\rangle$ vs $\epsilon_{\langle z^{\infty+2}\rangle}$ (solid curve) shows that the 2D outward Elsässer energy is proportional to $\epsilon_{\langle z^{\infty+2}\rangle}$. However, the theoretical $\langle z^{*+2}\rangle$ vs $\epsilon_{\langle z^{*+2}\rangle}$ (dashed curve) shows that they are initially negatively correlated, and then positively correlated. The theoretical $\langle z^{*+2}\rangle$ vs $\epsilon_{\langle z^{*+2}\rangle}$ and $\langle z^{*+2}\rangle$ vs $\epsilon_{\langle z^{\infty+2}\rangle}$ curves are close to the observed result (black triangle). In the right panel of Figure 5, the theoretical T vs $\epsilon_{\langle z^{\infty+2}\rangle}$ (solid curve) and the theoretical T

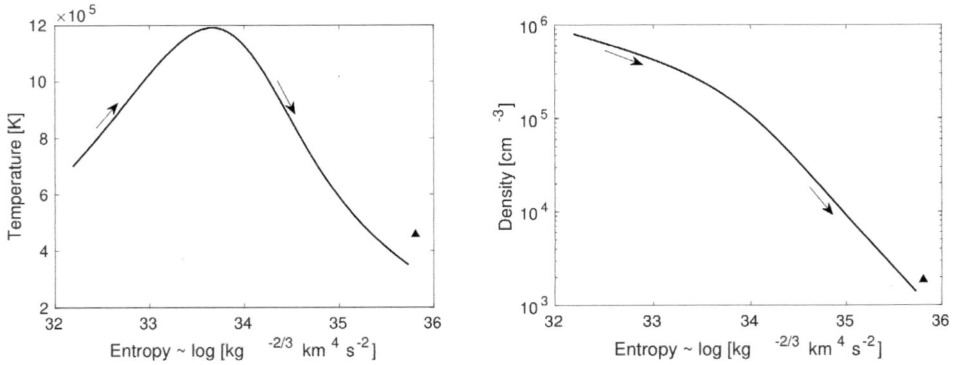

Figure 4. Left-panel: Proton temperature as a function of proton entropy. Right-panel: Proton density as a function of proton entropy. The format of the figure is similar to Figure 3.

vs $\epsilon_{\langle z^{*+2}\rangle}$ (dashed curve) show that at first T is inversely proportional to the 2D/slab heating rate, and then they are directly proportional. Finally, the bottom panel of Figure 5 shows that the solar wind speed is inversely proportional to $\epsilon_{\langle z^{\infty,*+2}\rangle}$, where the theoretical U vs $\epsilon_{\langle z^{*+2}\rangle}$ (dashed) curve steeper than the theoretical U vs $\epsilon_{\langle z^{\infty+2}\rangle}$ (solid) curve.

3. Discussion and Conclusions

Adhikari et al [6] compared the theoretical quasi-2D turbulence solar wind heating model with measurement derived from Metis/SolO – PSP quadrature investigated by Telloni et al [19]. They measured the Alfvénic slow solar wind plasma traveling from the extended solar corona (3.5 – 6.3 $R_\odot$) to the very inner heliosphere (23.2 $R_\odot$). The average speed of the slow solar wind measured by PSP is about 219.34 kms^{-1} and the normalized cross-helicity is about 0.96 [6,19]. In this manuscript, we study the evolution of proton entropy and the turbulent cascade rate of the outward Elsässer energy during the Metis/SolO – PSP quadrature, and explore their relationship with the solar wind speed, solar wind density, and solar wind temperature. We summarize our findings as follows.

(i) The theoretical solar wind proton entropy increases rapidly over a small height from 3.3 $R_\odot$ to $\sim$ 4 $R_\odot$, which may be related to the dissipation of 2D turbulence energy (e.g., Adhikari et al [24]), and then increases gradually with distance. The theoretical entropy is close to the observed entropy measured by PSP at 23.2 $R_\odot$.

(ii) The theoretical 2D heating rate of the outward Elsässer energy is larger than the corresponding theoretical slab heating rate in the extended solar corona. This indicates that the heating rate due to 2D turbulence is the main component for heating the solar corona.

(iii) Initially, the theoretical proton temperature is inversely proportional to the theoretical 2D outward Elsässer energy near 3.3 $R_\odot$, and then the two are

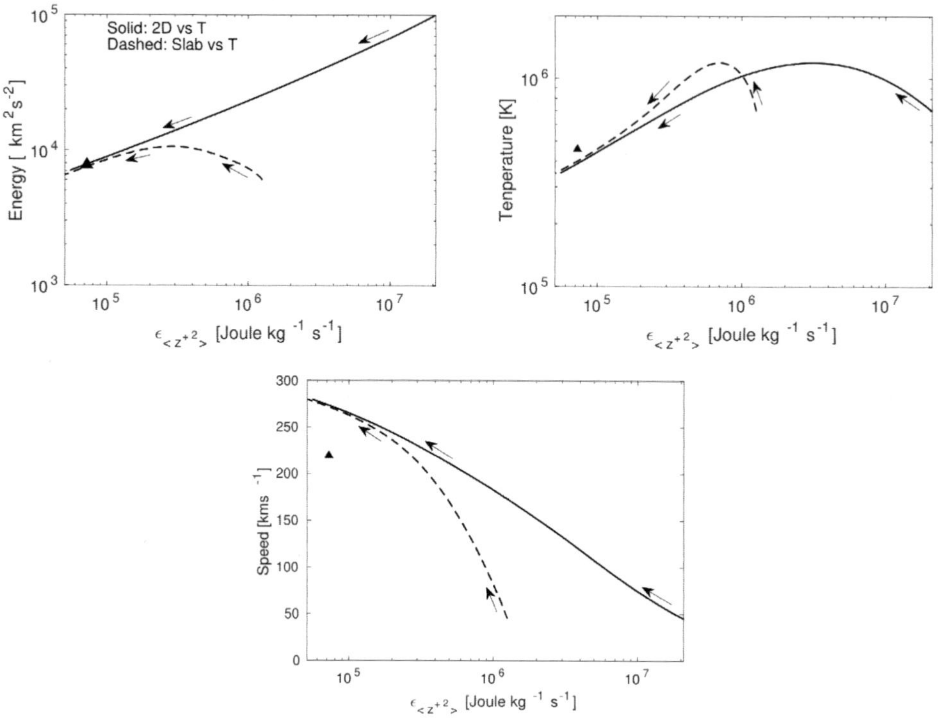

Figure 5. Top-left: Outward Elsässer energy as a function of turbulence cascade rate. Top-right: Proton temperature as a function of turbulence cascade rate. Bottom: Solar wind speed as a function of turbulence cascade rate. The format of the figure is similar to Figure 3.

proportional. Similarly, the theoretical 2D outward Elsässer energy is inversely proportional to the solar wind speed.

(iv) The theoretical proton temperature is proportional to the theoretical solar wind speed and the theoretical proton entropy until the proton temperature reaches a maximum value, and after which the temperature is inversely proportional to them. The theoretical proton density is inversely proportional to the theoretical proton entropy.

(v) The theoretical 2D outward Elsässer energy is proportional to the theoretical cascade rate of the 2D outward Elsässer energy.

Acknowledgments

We acknowledge the partial support of a Parker Solar Probe contract SV4-84017, an NSF EPSCoR RII-Track-1 cooperative agreement OIA-1655280, and NASA awards 80NSSC20K1783 and 80NSSC21K1319.

References

[1] Kasper J C, Klein K G, Lichko E et al 2021 *PhRvL* 127 255101
[2] Fox N J, Velli M C, Bale S D et al 2016 *SSRv* 204 7
[3] Zank G p, Zhao L L, Adhikari L, Telloni D et al 2022 *ApJ*, 926 9
[4] Antonucci E, Romoli M, Andretta V et al 2020 *A&A* 642 A10
[5] Telloni D, Zank G P, Sorriso-Valvo L et al 2022a *ApJ* 935 112
[6] Adhikari L, Zank G P, Telloni D, & Zhao L L et al 2022a *ApJL* 937 10
[7] Telloni D, Antonucci E, Adhikari L, Zank G P et al 2023 *A&A* 670 18
[8] Matthaeus W H, Zank G P, Oughton S, Mullan D J, & Dmitruk P 1999 *ApJL* 523 L93
[9] Zank G P, Adhikari L, Hunana P et al 2018 *ApJ* 854 32
[10] Adhikari L, Zank G P, & Zhao L L et al 2020a *ApJ* 901 14
[11] Bale S D, Goetz K, Harvey P R et al 2016, *SSRv* 204 49
[12] Kasper J C, Abiad R, Austin G et al 2016 *SSRv* 204 131
[13] Horbury T S, O'Brien H, Carrasco Blazquez I et al 2020, *A&A* 642 A9
[14] Owen C J, Bruno R, Livi S et al 2020 *A&A* 642 A16
[15] Adhikari L, Zank G P, Zhao L L, & Telloni D et al 2022b *ApJ* 933 56
[16] Adhikari L, Zank G P, Zhao L L, & Telloni D et al 2022c *ApJ* 938 12
[17] Zank G P, Zhao L L, Adhikari L et al 2021 *PhPl* 28 080501
[18] Telloni D, Sorriso-Valvo L, Woodham L D, Panasenco O et al 2021 *ApJL* 912 8
[19] Telloni D, Zank G P, Sorriso-Valvo L et al 2022b *ApJ* 935 112
[20] Panasenco O, Velli M, D'Amicis R et al 2020 *ApJS* 246 54
[21] Biondo R, Bemporad A, Pagano P, Telloni D et al A&A 668 5
[22] Whang Y C, Behannon K W, Burlaga L F, & Zhang S 1989 *JGR* 94 2345
[23] Whang Y C, Liu S, & Burlaga L F 1990 *JGR* 95 18769
[24] Adhikari L, Zank G P, Zhao L L, & Webb G M et al 2020b *ApJ* 891 7
[25] Zank G P, Adhikari L, Hunana P et al 2017 *ApJ* 835 147
[26] Adhikari L, Zank G P, & Zhao L 2021 *Fluid* 6 368
[27] Adhikari L, Zank G P, Telloni D et al 2017 *ApJ* 851 117
[28] Maksimovic M, Bale S D, Bercic L et al 2020 *ApJS* 246 62
[29] Halekas J S, Whittlesey P, Larson D E, Maksimovic E et al 2022 *ApJ* 936 53

On the Causality Problem in Focused Particle Transport

R D Strauss[1,2], J P van den Berg[1], N E Engelbrecht[1,2], N Wijsen[3,4]
[1]Center for Space Research, North-West University, 10 Hoffman street, Potchefstroom, 2532, South Africa
[2]National Institute for Theoretical and Computational Sciences (NITheCS), South Africa
[3]NASA, Goddard Space Flight Center, Heliophysics Science Division, USA
[4]Department of Astronomy, University of Maryland, College Park, MD 20742, USA

E-mail: `dutoit.strauss@nwu.ac.za`

Abstract. When perpendicular diffusion is included into the focused transport equation, certain assumptions of the perpendicular diffusion coefficient (including the well-known fieldline random walk limit) clearly lead to faster-than-light motion for relativistic particles, and transport that does not preserve causality in general. In this work we show that this discrepancy can be solved by including the effect of turbulence on particle drifts, especially the streaming term present in the focused transport equation, leading to a so-called drift reduction factor. We discuss this drift reduction factor, how it solves the above-mentioned causality problem, and its general implications for particle transport. Under certain simplifying assumptions we show the interplay between particle streaming (as a drift process) and perpendicular diffusion, and how these processes can be consistently coupled through the turbulent magnetic field.

1. Introduction

Normal diffusive models are notorious for not preserving causality. Brownian motion, for instance, broadens an initial narrow distribution into a Gaussian-like function, requiring, in principle, that some particles propagate with an infinite diffusive speed. In addition to this, numerical transport models sometimes struggle to preserve causality due to numerical truncation errors that tend to lead to numerical diffusion; an artifact that again does not preserve causality. While we acknowledge these difficulties in modelling particle transport, this paper discussed a more fundamental issue: That the parameters introduced into a particle transport model does not necessarily preserve causality when chosen arbitrarily and inconsistent with each other.

Fig. 1, taken from [1], shows simulations of 100 keV solar energetic particle (SEP) electrons in the equatorial plane of the heliosphere with pitch-angle and perpendicular diffusion included. The red dashed circle shows the position of a ballistic electron beam released at the same time as the simulated electrons. In panels (b) and (c) it is clear that some modelled electrons pass this causal barrier. These particles move (significantly) faster than their speed allow. While numerical and diffusive issues may contribute to this non-causal transport, [1] showed that this motion is an implication of the implemented transport coefficients that allow, for this case, particles (with a pitch-angle cosine of $\mu = 1$) to move a certain distance along the mean magnetic

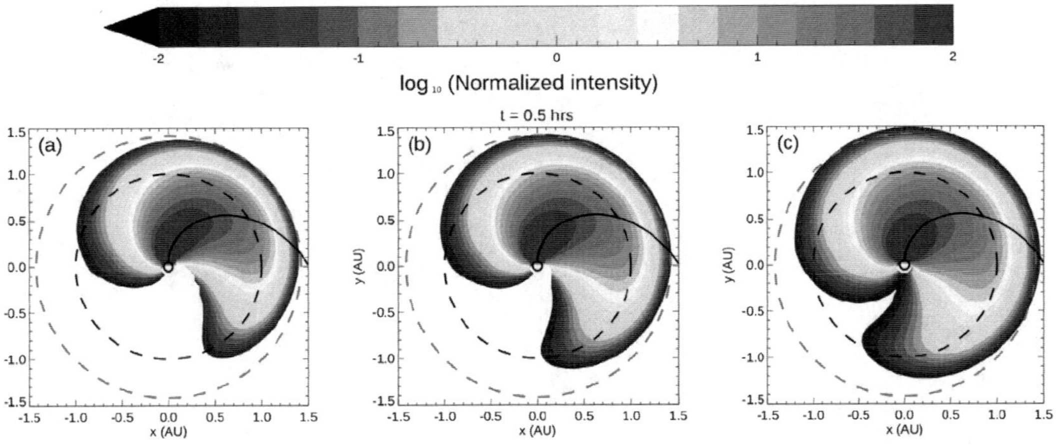

Figure 1. Simulations of the intensity of 100 keV electrons in the equatorial plane of the heliosphere for three different choices of the perpendicular diffusion coefficient (a: $D_\perp \propto \sqrt{1-\mu^2}$; b: $D_\perp \propto$ const; c: $D_\perp \propto |\mu|$). The figure is taken from [1] and shows the intensity 30 minutes after particle injection near the Sun.

field ($\Delta s_{||} = v\Delta t$) and, simultaneously, move a distance ($\Delta s_\perp \neq 0$) perpendicular to the mean field so that (see their Eq. 13)

$$\Delta s = \sqrt{(\Delta s_{||})^2 + (\Delta s_\perp)^2} > v\Delta t, \tag{1}$$

where v is the particle speed and Δs is the total particle displacement in a time increment Δt. These particles propagate faster than their speed allow and do not preserve causality.

One might argue that this is a limitation of the focused transport equation itself: The traditional derivation of this transport equation does not include the perpendicular diffusion process [e.g. 2]. However, as shown by e.g. [3], if the order of the averaging operations are chosen carefully when performing the gyro-averaging, the perpendicular diffusion terms emerge naturally. The inclusion of the perpendicular diffusion terms in the more general focused transport equation is therefore not an *ad-hoc* modification. It is also noteworthy to point out that the model of [4] that simulates meandering fieldlines does not have this causality problem.

This causality issue has mostly been ignored by the community (including by the present authors) but is addressed, and to some extent, solved in this proceedings paper.

2. Drift reduction in the presence of turbulence

Early galactic cosmic ray transport models showed that the weak scattering (i.e. calculated in the absence of turbulence) drift coefficient over-estimates drift effects (gradient, curvature, and neutral sheet drifts) and that the drift coefficient must be scaled down [e.g. 5]. While early work implemented various *ad hoc* forms of this suppression factor, full-orbit particle simulations by e.g. [6] showed that the presence of magnetic turbulence does indeed reduce the average drift speed by disrupting the drift process and that this reduction is physical and not a numerical artifact. [7] showed that a tractable analytical estimate of this factor can be obtained by using a simple perturbation approach. More recently, [8] extended this approach to the pitch-angle level, showing that drift effects, including the parallel streaming term present in the focused transport equation, are still suppressed in the presence of turbulence. In the next sections we show that

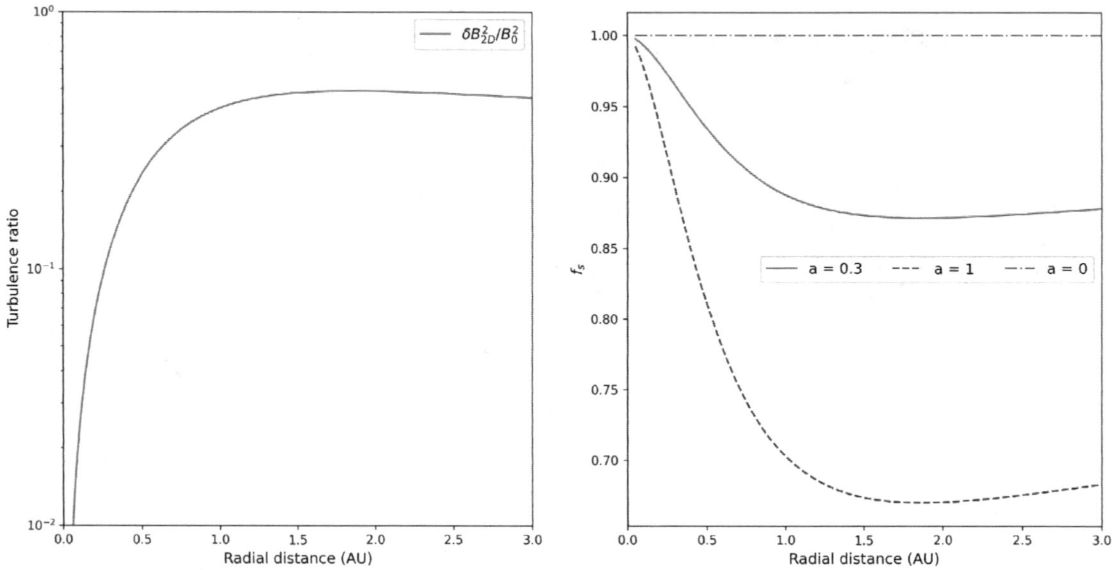

Figure 2. The left panel shows the assumed turbulence ratio from [10], while the right panel shows the corresponding drift suppression factor derived here for different values of a.

the inclusion of the drift suppression factor into the focused transport equation, in particular the parallel streaming term, can solve the apparent causality problem highlighted earlier.

3. Drift and diffusion in a turbulent field

3.1. Parallel transport (particle streaming)

Consider a weakly turbulent magnetic field that can be decomposed into a large-scale mean field $\vec{B_0}$ and a small fluctuating component, $\vec{B} = \vec{B_0} + \vec{\delta b}$. As usual, transverse turbulence is assumed, where $\vec{B_0} \cdot \vec{\delta b} = 0$, along with weak turbulence, $\vec{\delta b} \cdot \vec{\delta b} = \delta b^2 \ll B_0^2$, and after suitable averaging, $\langle \vec{\delta b} \rangle = \vec{0}$.

If $\vec{B_0}$ is in the $\hat{z}$ direction and $\vec{\delta b}$ in the $\hat{y}$ direction, the displacement of the guiding center (GC) is, after a time interval Δt,

$$\frac{\Delta y}{\Delta z} = a \frac{\delta b}{B_0} \tag{2}$$

where $a \in [0, 1]$ is introduced to give a probability of the GC following the turbulent field line, accounting for possible finite gyroradius effects. If pitch-angle scattering is, for the moment, neglected, then the GC moves a distance $(\Delta l)^2 = (\Delta y)^2 + (\Delta z)^2$ in a time interval Δt along the instantaneous field, $\vec{B}$, so that the GC velocity *along the instantaneous magnetic field* is

$$\left(\frac{\Delta l}{\Delta t}\right)^2 = \left(\frac{\Delta z}{\Delta t}\right)^2 \left[1 + a^2 \frac{\delta b^2}{B_0^2}\right] \tag{3}$$

If $v_{||} = \lim_{\Delta t \to 0} \Delta l / \Delta t$ is the GC velocity *along the instantaneous magnetic field* and $v_{0,||} = \lim_{\Delta t \to 0} \Delta z / \Delta t$ is the GC velocity *along the mean magnetic field* , then

$$v_{0,||} = \left[1 + a^2 \frac{\delta b^2}{B_0^2}\right]^{-1/2} v_{||} = \sqrt{f_s} v_{||} \tag{4}$$

where

$$f_s = \left[1 + a^2 \frac{\delta b^2}{B_0^2}\right]^{-1} \tag{5}$$

the so-called *drift suppression factor* using the notation of [8]. With the inclusion of transversal fluctuations, the GC can travel perpendicular to the mean field with the result that the GC motion along the mean field is reduced, i.e. $v_{0,||} < v_{||}$. Using the turbulence parameters as assumed by [10], this factor is shown in Fig. 2 as a function of radial distance.

3.2. Perpendicular transport (perpendicular particle diffusion)

Similarly, the speed component perpendicular *to the mean field* is

$$v_{0,\perp} = \lim_{\Delta t \to 0} \frac{\Delta y}{\Delta t} = \lim_{\Delta t \to 0} \frac{\Delta z}{\Delta t} \frac{\Delta y}{\Delta z} = a v_{0,||} \frac{\delta b}{B_0}, \tag{6}$$

which averages to $\langle v_{0,\perp} \rangle = 0$ as this is a purely stochastic motion. However, the second moment thereof, $\langle v_{0,\perp}^2 \rangle$, gives rise to perpendicular diffusion. Following the same steps as in [9], we can obtain the fieldline random walk (FLRW) diffusion coefficient as

$$D_\perp = \frac{1}{2} \int_0^\infty \langle v_{0\perp}(0) v_{0,\perp}(t) \rangle dt = \langle v_{0,\perp}^2(0) \rangle \int_0^\infty \exp\left(-\frac{t}{\tau_{\text{dec}}}\right) dt = \langle v_{0,\perp}^2 \rangle \tau_{\text{dec}} \tag{7}$$

where exponential temporal decorrelation on a timescale of τ_{dec} was assumed. If we furthermore assume particles decorrelate by moving a perpendicular correlation length perpendicular to the mean field,

$$\tau_{\text{dec}} = \sqrt{2 \langle l_\perp^2 \rangle / \langle v_{0,\perp}^2 \rangle} \tag{8}$$

and we end up with

$$D_\perp = a \left| v_{0,||} \right| \sqrt{\langle l_\perp^2 \rangle \frac{\delta b^2}{2 B_0^2}}. \tag{9}$$

where Eq. 6 was substituted for the perpendicular speed. If we assume $\sqrt{\langle l_\perp^2 \rangle}$ is given by the so-called turbulence ultrascale, λ_u [see 10, for additional details], we end up with

$$D_\perp = a \left| v_{0,||} \right| \kappa = a \sqrt{f_s} |\mu| v \kappa, \tag{10}$$

where $\kappa = \lambda_u \sqrt{\delta b^2 / 2 B_0^2}$ is the so-called FLRW diffusion coefficient which described the diffusion of the magnetic field lines.

3.3. A causality test

The treatment above solves this causality issue via a drift reduction factor that is self consistently coupled to a perpendicular diffusion coefficient by means of basic turbulence quantities. This is illustrated by considering the following limits: If $\delta b \to 0$, particles stream along the mean field with speed μv as $f_s \to 1$, while, simultaneously, $D_\perp \to 0$. As δb is then increased, the parallel streaming decreases while $D_\perp$ increases; the perpendicular motion (through diffusion) is obtained by decreasing the parallel motion (i.e. particle streaming). These coefficients also change consistently when a is changed.

Journal of Physics: Conference Series **2544** (2023) 012008 doi:10.1088/1742-6596/2544/1/012008

Within the revised FLRW framework presented here, the kinetic energy associated with parallel motion is converted into perpendicular motion due to GCs following turbulent magnetic field lines, i.e. the total kinetic energy associated with the motion of the GC is conserved,

$$\langle v_{0,||}^2 \rangle + \langle v_{0,\perp}^2 \rangle = \mu^2 v^2 = v_{||}^2 \tag{11}$$

where $v_{||}$ is the parallel speed of the GC with respect to the instantaneous field and is a constant of motion if pitch-angle diffusion is neglected. *If the f_s factor is not included, perpendicular motion (kinetic energy) is introduced into the system violating the conservation of kinetic energy and leading to the causality violation introduced at the beginning of this paper.*

4. Example of causally correct model

Consider the following focused transport equation which contains perpendicular diffusion,

$$\frac{\partial f}{\partial t} + \nabla \cdot (\vec{v}_{gc} f) + \frac{\partial}{\partial \mu}\left(\frac{1-\mu^2}{2L} v f\right) = \frac{\partial}{\partial \mu}\left(D_{\mu\mu}\frac{\partial f}{\partial \mu}\right) + \nabla \cdot \left(\mathbf{D}_{\perp}^{(x)} \cdot \nabla f\right). \tag{12}$$

While the pitch-angle diffusion coefficient $D_{\mu\mu}$ can be calculated from the slab component of the turbulence [see the discussion by 10], the effect of the 2D component (which includes transversal complexity) is now considered. When performing a first-order perturbation of the GC velocity in a turbulent field, [8] show that the resulting GC velocity can be decomposed into a deterministic and diffusive component, $\vec{v}_{gc} \to \vec{v}_{0,gc} + \delta\vec{v}_{gc}$ where, after appropriate averaging, $\langle \delta\vec{v}_{gc} \rangle = \vec{0}$. The effect of turbulence is, however, still present in the deterministic term. The streaming term, for example, becomes,

$$\vec{v}_{0,gc} = \frac{dz}{dt}\hat{b} = \left[1 + a^2\frac{\delta b^2}{B_0^2}\right]^{-1/2} v_{||}\hat{b} = \sqrt{f_s} v_{||}\hat{b} \tag{13}$$

where $\hat{b} = \vec{B}_0/B_0$ is a unit vector along the mean magnetic field and f_s is the drift suppression factor given by Eq. 5. It is furthermore assumed that only the 2D turbulence component influences drift effects and that the total energy in this turbulent field is given by

$$\delta b^2 = 2\pi \int_0^\infty g_{2D}(k_\perp) dk_\perp, \tag{14}$$

where g_{2D} denotes the 2D turbulence power spectrum. The stochastic drift motion is now added into the calculation of the perpendicular diffusion coefficient [9],

$$D_\perp = a\sqrt{f_s}|\mu| v \kappa \tag{15}$$

with

$$\kappa^2 = \frac{\pi}{B_0^2}\int_0^\infty k_\perp^{-2} g_{2D}(k_\perp) dk_\perp. \tag{16}$$

This streaming velocity and coupled perpendicular diffusion coefficient can now be included in the focused transport equation. There is now an interplay between the deterministic drift motion and the diffusion transverse motion: In the limit of vanishing fluctuations, $\delta b^2 \to 0$, the streaming term approaches the traditional weak-scattering form, but perpendicular diffusion also disappears. With turbulence included in the system, turbulent motion is included which leads to perpendicular diffusion. This stochastic motion occurs, however, at the expense of the deterministic motion and visa versa. *It is therefore not possible to include perpendicular diffusion without sacrificing drift motion. You cannot have your cake and eat it.*

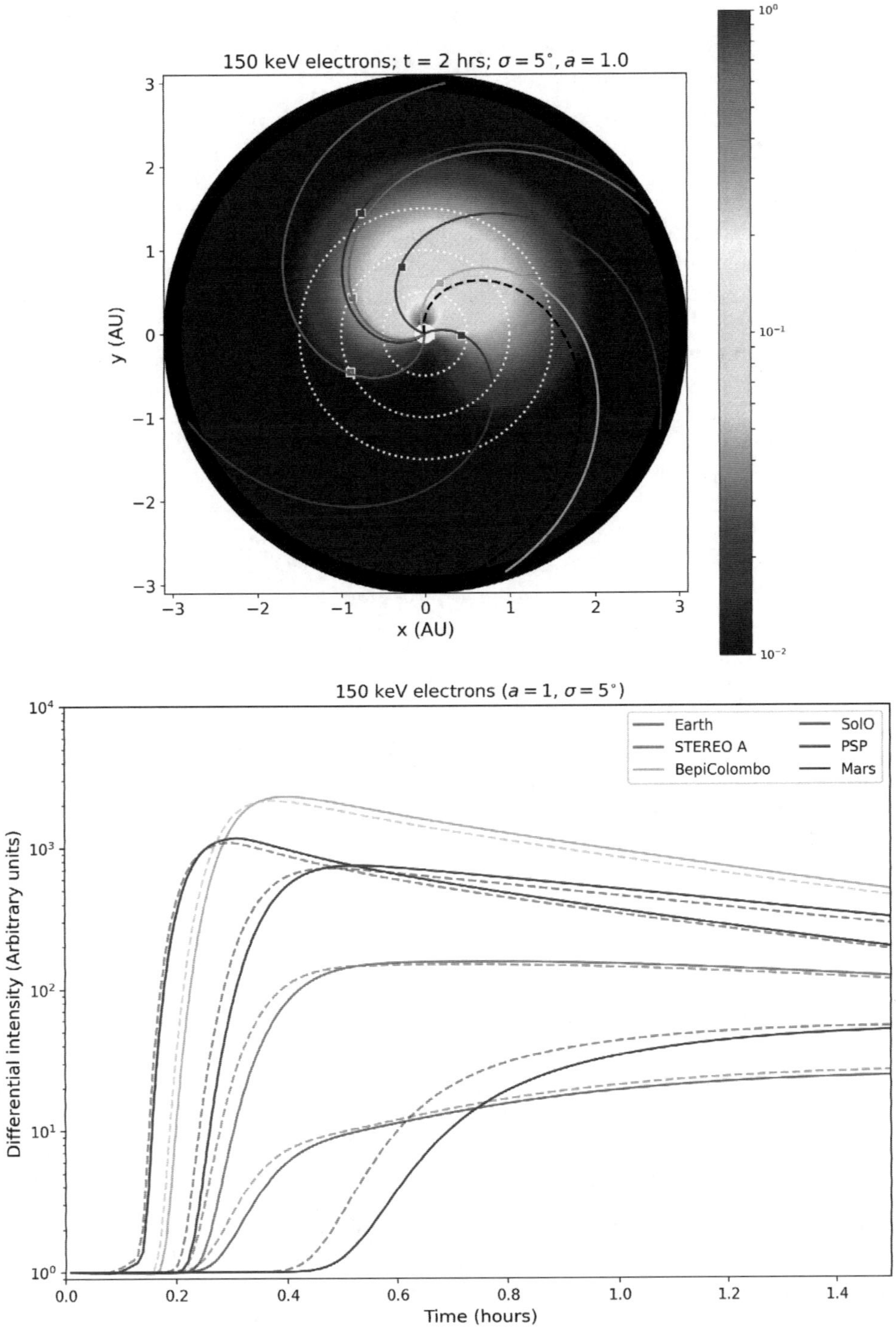

Figure 3. Simulations results for 150 keV SEP electrons using the model of [10] and a maximum value of $a = 1$. The top panel shows the intensity in the equatorial plane, with the nominal magnetic connectivity to a number of virtual spacecraft. The bottom panel shows the resulting differential intensity at these observers as a function of time. The dashed curves are for simulations that ignore the streaming reduction term.

The previously neglected reduction in the streaming term is now included in the model of [10] and compared in the bottom panel of Fig. 3 to the case without this correction (the dashed curves). Simulation results are shown at a number of virtual spacecraft (their positions indicated in the top panel).

5. Summary

In this short proceedings paper we have emphasized the causality problem that exists in the focused transport equation when including perpendicular diffusion. We propose that this causality problem can be solved by including the effects of turbulence on the deterministic drift terms, leading to reduced particle streaming along the mean magnetic field for the scenario considered here. Essentially, the causality problem is present because inconsistent coefficients are implemented in the transport equation: When 2D turbulence is introduced it will, simultaneously, lead to perpendicular diffusion and reduce the parallel streaming term. If these coefficients are chosen arbitrarily, e.g. including perpendicular diffusion and adopting the weak-scattering streaming (drift) term, causality errors are present in the model.

The difficulty still lies in calculating the transport coefficients self-consistently. Here, we have made use of a simplified perturbation approach to derive the drift suppression factor and the FLRW perpendicular diffusion coefficient. These coefficients are inherently coupled and a consistent pair must be incorporated into transport models. The derivation of the drift reduction factor presented here, although conceptually interesting, is very simplistic. Extensions beyond the simplified assumptions presented here are indeed possible [see 8] and can potentially lead to even larger streaming reduction effects. In addition, the approach followed here predicts that $\langle v_{0,||}^2 \rangle \neq 0$ which implies that parallel spatial diffusion should also be present in the focused transport equation. It is not clear if this is a robust result, given the basic set-up assumed here.

The effect of the drift suppression factor (i.e. the reduced streaming factor) on simulated particle intensities is yet to be explored in detail, with preliminary simulation results presented in Fig. 3. Generally, with the drift velocity along the mean magnetic field reduced, particles will take longer to reach any observational point in the inner heliosphere. This, in turn, will lead to a longer onset time for SEP events and longer deduced magnetic field pathlengths, similar to the findings of [11].

References

[1] Strauss R D and Fichtner H 2015 *Astrophys. J.* **801** 29

[2] Skilling J 1971 *Astrophys. J.* **170** 265

[3] Zhang M J 2006 *J. Geophys. R.* **111** A04208

[4] Laitinen T and Dalla S 2019 *Astrophys. J.* **887** 222

[5] Potgieter M S and Burger R A 1990 *Astron. Astrophys.* **233** 598

[6] Minnie J, Bieber J W, Matthaeus W H and Burger R A 2007 *Astrophys. J.* **670** 1149

[7] Engelbrecht N E, Strauss R D, le Roux J A and Burger R A 2017 *Astrophys. J.* **841** 107

[8] van den Berg J P, Engelbrecht N E, Wijsen N and Strauss R D 2021 *Astrophys. J.* **922** 200

[9] Strauss R D, le Roux J A, Engelbrecht N E, Ruffolo D and Dunzlaff P 2016 *Astrophys. J.* **825** 43

[10] Strauss R D, Dresing N and Engelbrecht N E 2017 *Astrophys. J.* **837** 43

[11] Chhiber R et al. 2021 *Astron. Astrophys.* **650** A26

Transmission of magnetic island modes across interplanetary shocks: comparison of theory and observations

A Pitna[1], G P Zank[2], M Nakanotani[2], L-L Zhao[2], L Adhikari[2], J Safrankova[1] and Z Nemecek[1]

[1] Charles University, Faculty of Mathematics and Physics, V Holesovickach 2, Prague, CZ-18000

[2] Center for Space Plasma and Aeronomic Research (CSPAR), University of Alabama in Huntsville, Huntsville, AL 35899, USA

E-mail: alex@aurora.troja.mff.cuni.cz

Abstract. Interplanetary shock waves are observed frequently in turbulent solar wind. They naturally enhance the temperature/entropy of the plasma through which they propagate. Moreover, many studies have shown that they also act as an amplifier of the fluctuations incident on the shock front. Solar wind turbulent fluctuations can be well described as the superposition of quasi-2D and slab components, the former being energetically dominant. In this paper, we address the interaction of fast forward shocks observed by the Wind spacecraft at 1 AU and quasi-2D turbulent fluctuations in the framework of the Zank et al. (2021) transmission model and we compare model predictions with observations. Our statistical study includes 378 shocks with varying upstream conditions and Mach numbers. We estimate the average ratio of the downstream observed and theoretically predicted power spectra within the inertial range of turbulence. We find that the distributions of this ratio for the whole set and for the subset of shocks that met the assumptions of the model, are remarkably close. We argue that a large statistical spread of the distributions of this ratio is governed by the inherent variation of the upstream conditions. Our findings suggest that the model predicts the downstream fluctuations with a good accuracy and that it may be adopted for a wider class of shocks than it was originally meant for.

1. Introduction

One of the outstanding open question of heliospheric physics that has recently attracted attention is the interaction of magnetohydrodynamic (MHD) turbulence and interplanetary (IP) shocks. The difficulty in solving this question lies in the complexity of the interaction, i.e., the character of the incident fluctuations, upstream bulk plasma parameters, the type of an IP shock, its obliquity, Mach number, compression ratio, shock evolution, etc., since all of these factors influence the resulting downstream state of the plasma. Thus, any dedicated study can only address a part of the questions.

Knowledge of how the upstream fluctuations are transmitted through the IP shock front provide insight into particle acceleration mechanisms, since the inhomogeneities in the upstream and downstream plasma play a crucial role in such processes [1]. The coupling of the solar wind plasma and Earth's magnetosphere is dependent on the level of the magnetic field fluctuations [2], thus, the enhanced downstream fluctuations may intensify this coupling, but a full understanding of the underlying physical mechanisms is missing.

It has been known for decades that IP shocks increase the levels of the incident/upstream fluctuations downstream [see e.g., review [3] and the references therein]. Most studies analyzed magnetic field fluctuations ([4, 5, 6, 7, 8, 9]). Recent investigations focus on the transmission of density [10], solar wind bulk velocity [9] and entropy fluctuations [11]. The transmission of Alfven waves was studied from a theoretical perspective by McKenzie and Westphal [12]. Adhikari et al. (2016) [7] studied the transport of incompressible turbulent fluctuations across quasi-parallel and quasi-perpendicular IP shocks using an incompressible MHD formulation of a turbulent transport model [13]. They reported that the turbulent energy increases across the shock and that the correlation length is approximately conserved across the shock.

Zank et al. (2021) [11] (hereafter Zank21) introduced a novel framework in which the transmission of MHD turbulent fluctuations can be straightforwardly investigated. The formulation was motivated by the observations that (1) the solar wind turbulent fluctuations can be successfully described as a superposition of an energetically dominant 2D and a minority slab component [14, 15] and (2) the majority of fast forward (FF) IP shocks are quasi-perpendicular [16], i.e., the angle between the shock normal, $\boldsymbol{n}$, and upstream magnetic field, $\boldsymbol{B}_{\mathrm{u}}$, θ_{Bn}, is larger than 45°. The model describes the linear interaction of incident vortical, magnetic island, entropy and acoustic normal modes with planar quasi-perpendicular shock. It assumes (1) a high upstream plasma beta, $\beta_{\mathrm{p}}^{\mathrm{u}} \gtrsim 1$, (the ratio of the plasma and magnetic pressure) and (2) strong magnetic field in a sense that the magnetic fluctuations, δB, are of the same order as the average B, $\delta B/B \approx 1$, (3) that the incident plasma velocity vector lies in the plane perpendicular to the average upstream magnetic field, and (4) wave vectors of the fluctuations are constrained in the same plane. Finally, Zank21 derived a set of linearized higher-order jump conditions for the velocity, density, pressure and magnetic field fluctuations. The jump equations for the magnetic field fluctuations read as

$$[U_{\mathrm{x}}\delta B_{\mathrm{y}}] = 0; \tag{1}$$

$$[\delta B_{\mathrm{x}}] = 0, \tag{2}$$

where square brackets denote the difference of the quantity from upstream to downstream, U_{x} is the normal component of upstream velocity in the shock frame, δB_{x} and δB_{y} are the normal and perpendicular components of the magnetic field fluctuation vector $\delta\boldsymbol{B}$. These equations are fully decoupled from the gas dynamic counterparts and thus can be solved separately.

Since the wave vector of the magnetic island mode is perpendicular to the average $\boldsymbol{B}$, this mode is non-propagating and it characterizes magnetic island/flux rope phenomenon [17, 18]. Adopting the NI turbulence approximation, magnetic field fluctuations in the upstream solar wind plasma can be viewed as a nonlinearly interacting set of the energetically dominant magnetic island modes. Solving eqs. (1) and (2) for incident magnetic island modes with upstream wave number k_1 and amplitude δB_1 yields the following solutions for the transmitted wave number k_2 and amplitude δB_2,

$$\tan\theta_2 = R^{-1}\tan\theta_1; \tag{3}$$

$$k_2 = R\frac{\cos\theta_1}{\cos\theta_2}k_1; \tag{4}$$

$$\delta B_2 = R\frac{\cos\theta_1}{\cos\theta_2}\delta B_1, \tag{5}$$

where θ_1 and θ_2 denote the upstream and downstream angles between the corresponding wave vector and shock normal, respectively, and R is the shock compression ratio.

In this paper, we analyze *in situ* measurements of a large set (378) of IP shocks observed by the Wind spacecraft with the aim of testing the theoretical basis for the transmission of MHD fluctuations introduced by Zank21. We analyze one hour intervals in both upstream and downstream regions. By employing a continuous wavelet transform (CWT), we estimate the trace power spectral densities (PSDs) of the magnetic field fluctuations. Based on the upstream PSDs of turbulent fluctuations, we estimate the theoretically predicted downstream PSDs and compare them with the observed downstream PSDs. We compute the average ratio of the observed and theoretically predicted PSDs within the inertial range for each case and we show that for a subset of IP shocks with high plasma beta, $\beta_{\mathrm{p}}^{\mathrm{u}} > 2$ and large obliquities, $\theta_{Bn} > 70°$, the median value of this ratio is ~ 0.73. We argue that the spread of the distribution of this ratio is given mainly by the variable upstream conditions and not due to the uncertainty of the model. Furthermore, we suggest that the Zank21 model can be useful also for a wider range of IP shocks than it was originally derived for.

2. Data analysis

Our statistical study is based on the IP shock observations made by the Wind spacecraft. We utilized the Heliospheric Shock Database generated at the University of Helsinki (http://ipshocks.fi/database) and we analyzed data from the Solar Wind Experiment (SWE) and Magnetic Field Investigation (MFI) instruments on board Wind [19, 20]. Our set includes IP shocks detected between December 1994 and May 2017. We excluded IP shocks that contain a large number of data gaps in both upstream and downstream one hour intervals (with less than 90% data coverage for MFI and 70% for SWE). The final data set includes 378 fast forward (FF) IP shocks. In Figure 1, we show the scatter plot of the angle, θ_{Bn} vs. upstream plasma beta, $\beta_{\mathrm{p}}^{\mathrm{u}}$ for the whole statistical set. Additionally, we define a subset (52 cases) of highly perpendicular IP shocks with large upstream beta, i.e., $\theta_{Bn} > 70°$ and $\beta_{\mathrm{p}}^{\mathrm{u}} > 2$, marked by the green diamonds. For the shocks in this subset, we calculate the level of the normalized magnetic field fluctuations within one hour, $\delta B/B_0$, in the upstream and downstream regions. Their mutual relation is shown in 2.

To determine the magnetic field power spectra in each one hour interval, we employ the widely used CWT [21, 22]. We estimate the trace power spectrum, P_B, and we denote the upstream and downstream PSDs as $P_B^{\mathrm{u}} = P_B^{\mathrm{u}}(k)$ and $P_B^{\mathrm{d}} = P_B^{\mathrm{d}}(k)$, respectively, where k denotes the wave number. We employed the Taylor hypothesis, i.e., we estimate k from the spacecraft frame frequencies f as $k = 2\pi f/v_{\mathrm{sw}}$, where v_{sw} denotes the average solar wind speed within the interval.

In calculations of the theoretical spectra, we used the same methodology as Zank21. We assumed that the underlying 2D upstream power spectrum is isotropic. We can then calculate the source terms for the upstream magnetic field fluctuations and we can use eqs. (3) to (5) to determine the downstream angle of a wave vector with respect to the shock normal, downstream wave number and magnetic field fluctuation amplitude. We estimate the power spectrum of the downstream fluctuations in k-θ_2 space and we integrate over θ_2 in order to obtain the downstream theoretical omni-directional spectrum, $P_B^{\mathrm{d,Th}}$.

3. Comparison of experimental and theoretical PSDs

In Figure 3, we present an example of experimental upstream/downstream power spectra and theoretical downstream power spectra for an IP shock observed by the Wind spacecraft on May 7, 2007 at 07:03 UT. We see that the agreement between P_B^{d} and $P_B^{\mathrm{d,th}}$ in the inertial range is very good. We quantify the similarity between the observed and theoretically predicted spectrum by calculating the average ratio between these spectra in a wave number range $k \in \left\langle 10^{-4}, 10^{-3} \right\rangle \mathrm{km}^{-1}$, denoted as $R_{\mathrm{OT}} = \left\langle P_B^{\mathrm{d}}/P_B^{\mathrm{d,Th}} \right\rangle$. We choose this range because it

Journal of Physics: Conference Series **2544** (2023) 012009 doi:10.1088/1742-6596/2544/1/012009

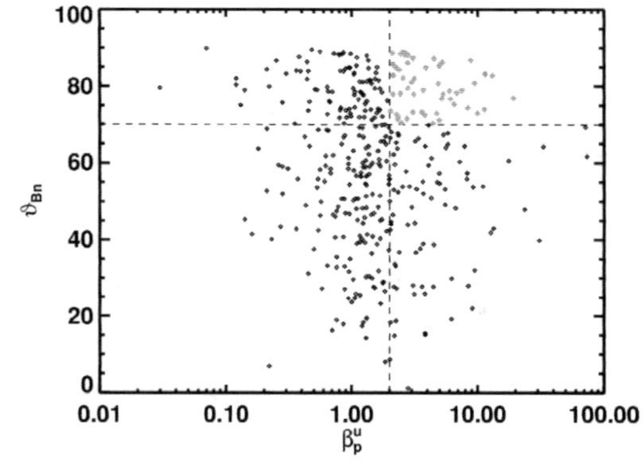

Figure 1. Diamonds show the variation of angle θ_{Bn} between the shock normal and upstream magnetic field as a function of upstream proton plasma $\beta_{\rm p}^{\rm u}$ for the whole set of 378 IP shocks. Green diamonds identify the sub-sample with $\beta_{\rm p}^{\rm u} > 2$ and $\theta_{Bn} > 70°$.

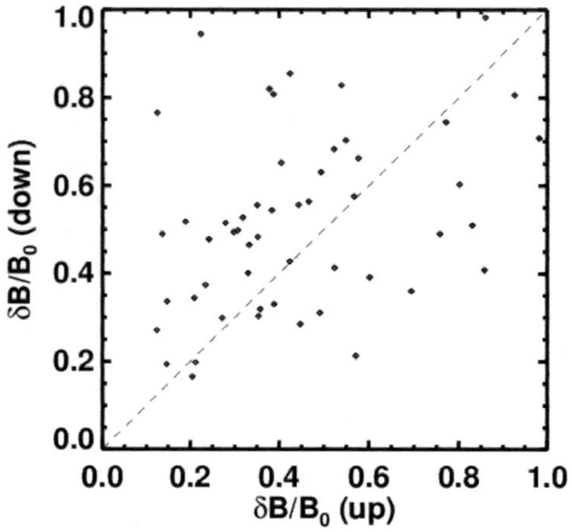

Figure 2. Scatter plot of downstream normalized magnetic field fluctuations versus the corresponding upstream values for the subset.

avoids PSD values that either (1) exhibit large errors due to statistical under-sampling for low k and (2) are sufficiently above the characteristic kinetic scales (high k).

We estimate $R_{\rm OT}$ for each case in the subset of quasi-perpendicular high beta IP shocks. Figure 4a shows the distribution of $R_{\rm OT}$. The median value of the distribution is $0.73^{+0.05}_{-0.09}$, signifying that the theoretical PSDs exhibit slightly higher values than the corresponding observed PSDs. On the other hand, the distribution itself is highly non-gaussian with a large variance. In order to quantify the spread of the distribution, we employ an estimator that is less sensitive to outliers, a so-called central confidence interval factor (CIF) [23], which is defined as the ratio of the upper and lower values of the central confidence interval (CI). We choose a CI that contains 68.3% of all values. The calculation yields $CIF(R_{\rm OT}) = 6.7^{+2.8}_{-0.9}$. Roughly, this number can be viewed as an 'uncertainty factor' between the observed and modeled quantity. One can say that the model differs from the observation by a factor of $CIF(R_{\rm OT})/2$, on average. Three main factors that govern the value of CIF are (1) the model uncertainty, (2) uncertainty in the levels of upstream fluctuations and (3) uncertainty in compression ratio. We address these aspects in the following paragraphs.

Addressing the model accuracy, we inspect the dependence of upstream $\beta_{\rm p}^{\rm u}$, θ_{Bn} and $\delta B/B_0$ as a function of $R_{\rm OT}$, in Figures 4b, 4c and 4d, respectively. The expectation from the Zank21 theory would be that for high $\beta_{\rm p}^{\rm u}$, high θ_{Bn} and/or large $\delta B/B_0$, the values of $R_{\rm OT}$ should cluster around 1 and for lower $\beta_{\rm p}^{\rm u}$, θ_{Bn} and $\delta B/B_0$, one would expect a deviation from unity. However, we see no clearly distinguishable trend in these figures. Due to the absence of the anticipated deviations in $R_{\rm OT}$, we analyze all cases (378) in the same manner.

We estimate $R_{\rm OT}$ for the whole statistical sample and we plot its distribution in Figure 5a. We see that the parameters of the distribution, $median(R_{\rm OT}) = 0.63^{+0.06}_{-0.07}$ and $CIF(R_{\rm OT}) = 8.0^{+0.9}_{-0.7}$, are remarkably close to those of the sub-sample. It is reasonable to expect that the median value of the whole sample should differ from the median value of the sub-sample which satisfies the assumptions of the theory, and moreover, the CIF should exhibit larger values for the whole

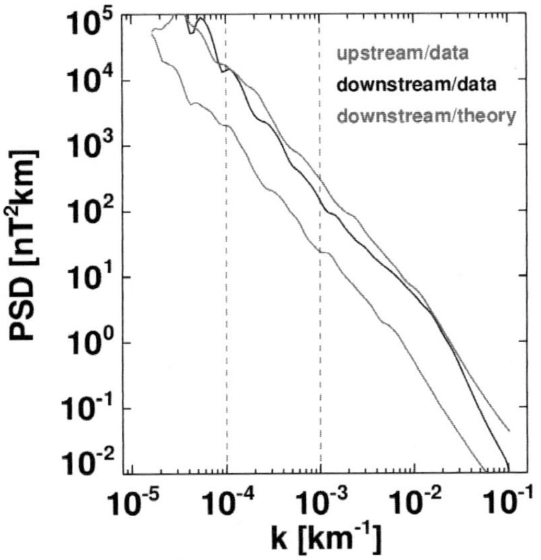

Figure 3. Magnetic field power spectra upstream (blue) and downstream (black) of IP shock front. The red curve shows the theoretically predicted power spectrum. Dashed vertical lines mark the wave number range within which the average of the ratio between the observed and theoretically predicted downstream spectra are computed. For this case, $R_{\mathrm{OT}} = 0.65$.

statistical set. We observe this trend, but the uncertainties[1] prohibit a firm conclusion about its significance.

The dependencies of R_{OT} on $\beta_{\mathrm{p}}^{\mathrm{u}}$ and θ_{Bn} for the whole set show no correlations between these quantities, as Figures 5b and 5d demonstrate. On the other hand, Figure 5d shows $\delta B/B_0$ as a function of R_{OT} and we see that for $\delta B/B_0 \lesssim 0.2$, R_{OT} tends to be higher than 1, which suggests that the theory is not applicable for very low levels of upstream fluctuations. This is not surprising, since it is one of the assumptions of the theory. However, we have now obtained an empirical value for a lower acceptable limit of $\delta B/B_0$.

We also investigate the uncertainty in P_B^{u}. The "uncertainties" in the levels of upstream fluctuations can be understood in the following sense: the original levels of fluctuations that were transmitted through the shock front and which give rise to the downstream observed fluctuations (P_B^{d}) are not the same as the fluctuations actually measured in the upstream region later (P_B^{u}). Let us denote the PSD of these fluctuations as $P_B^{\mathrm{u,ori}}$. The key question that has to be addressed is: what is the distribution of the ratio $\left\langle P_B^{\mathrm{u,ori}}/P_B^{\mathrm{u}}\right\rangle$? It is obviously impossible to estimate this ratio from single point measurements for any particular IP shock. However, an excellent proxy for this distribution is statistics of the ratio of PSDs of two consecutive 1-hour intervals in the pristine solar wind.

Since R_{OT} is defined as the average in the specific k-range, we estimate the relevant

[1] In the estimation of the confidence intervals for median and CIF, we employ a bootstrap re-sampling method [24].

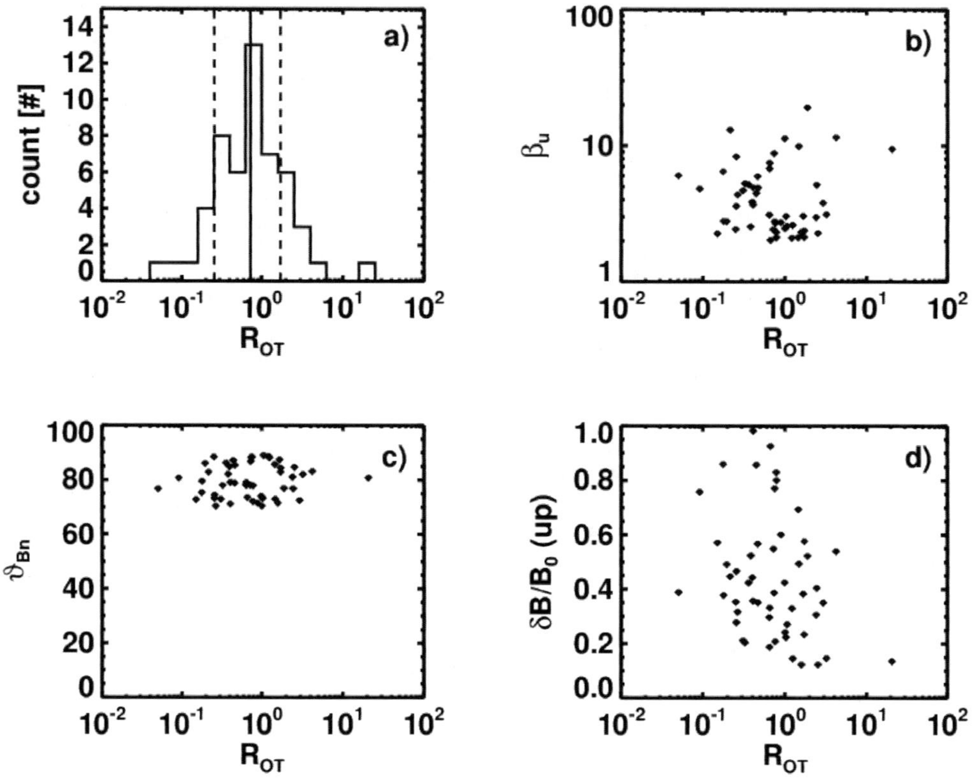

Figure 4. (a) Distribution of R_{OT} for the high proton plasma beta quasi-$\perp$ subset. Dashed vertical lines identify the central confident interval containing 68.3 % of data. Scatter plots on panels (b), (c) and (d) show the variation of $\beta_{\mathrm{p}}^{\mathrm{u}}$, θ_{Bn} and upstream $\delta B/B_0$ versus R_{OT}, respectively.

distribution as it follows: (1) We select the year of 2015 and estimate PSDs within all one-hour non-overlapping intervals. After applying the same selection criteria for the intervals (excluding intervals with large portions of data gaps), we obtain 6477 cases (out of 8760 possible), (2) We calculate the average ratio of PSDs, $R_{\mathrm{PSD\,B}}$, between all pairs of two consecutive intervals. The resulting distribution of $R_{\mathrm{PSD\,B}}$ is shown in Figure 6. Its median value is essentially one and its $CIF = 3.73^{+0.05}_{-0.09}$, which is less than $CIF(R_{\mathrm{OT}})$ for the subset, as might be expected. On the basis of this CIF value, we assert that roughly 50 % of the width of the distribution of R_{OT} (in terms of CIF) is given by the variability of the upstream plasma.

4. Discussion and conclusions

In this paper, we investigated the transmission of MHD scale fluctuations using the theoretical approach introduced by Zank21. Zank21 showed that for three particular cases observed by Wind, Ulysses and Voyager 2 at 1, 5 and 84 AU, respectively, the theoretically predicted power spectra match the observed counterparts very well. Here, we extended their study and we apply their methodology to a larger set of IP shocks observed by Wind at L1.

Perhaps the most surprising result is that the distributions of the ratios of observed and theoretically predicted PSD for the whole statistical set and the sub-sample (high beta, quasi-

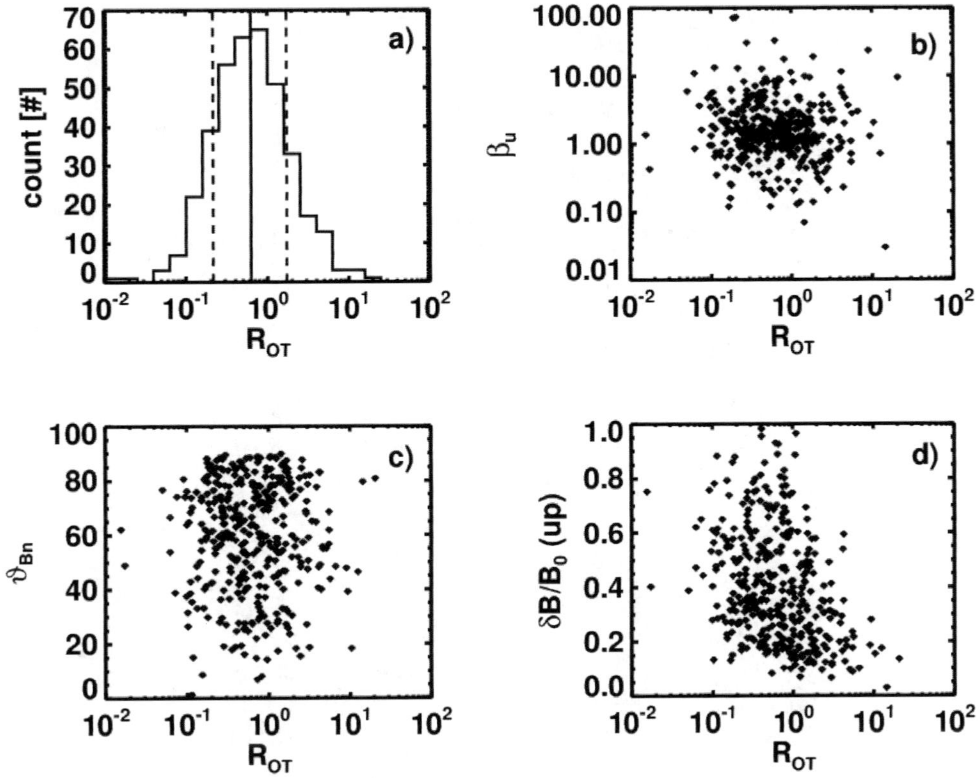

Figure 5. Distribution of R_{OT} for the whole statistical sample and the corresponding scatter plots. The format is the same as in Figure 4.

$\perp$ IP shocks) are remarkably similar. When we realize that 2D/magnetic island fluctuations dominate the turbulent energy budget in the solar wind, and majority of IP shocks exhibit quasi-perpendicular geometry, it can be then anticipated that we observe only small differences between the distributions. The distribution of $R_{\mathrm{PSD\,B}}$ demonstrates that the largest factor that influences the distribution of R_{OT} is the variable level of upstream fluctuations over the scales of interest. Moreover, the compression ratio plays a pivotal role in the model and any uncertainty in its estimation translates into uncertainty of the estimated R_{OT}. A quantitative assessment of the interplay between model and data uncertainties is not straightforward. However, a comparison of the CIFs of R_{PSDB} and R_{OT} suggests that the uncertainty is rather small.

The most obvious way to correct for the variability of the upstream solar wind conditions is to utilize multi-point spacecraft observations of the same IP shock. This can be achieved by either (1) radial alignment of sufficiently separated spacecraft (see, e.g., [25, 26]) or (2) by exploiting a standing bow shock (BS) (e.g., the terrestrial BS), where concurrent upstream and downstream measurements are commonplace (see, e.g., [27]). In both cases, the information about the incident fluctuations will be available and the source terms that enter the calculation (eqs. (3) to (5)) will be known with greater accuracy.

Numerical simulations of IP shocks propagating into magnetized turbulent plasma can provide a direct test of any turbulent transmission model. Indeed, Nakanotani et al. (2022) [28] ran a 2D hybrid kinetic simulation of a moderately strong shock (sonic Mach number of 2.4) propagating

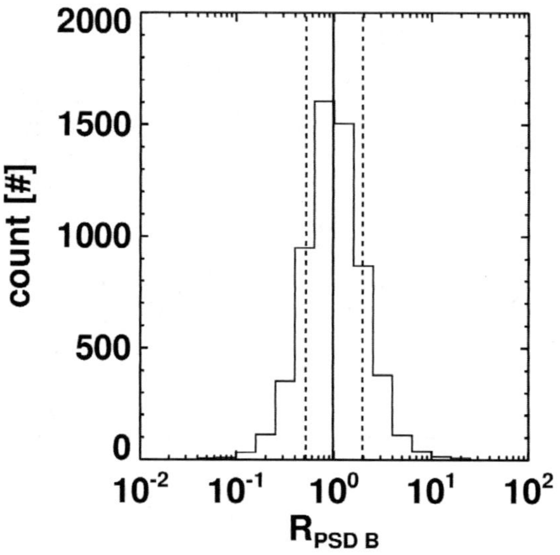

Figure 6. Distribution of the average ratio R_{PSDB} of power spectra estimated within two consecutive one hour long intervals of the solar wind (during year 2015). Dashed vertical lines identify the central confidence interval CI containing 68.3 % of data, whereas the solid vertical line identifies the median of the distribution.

into a turbulent plasma initiated by magnetic reconnection of alternating force-free magnetic field. They observed an enhancement of the levels of magnetic field fluctuations across the shock front that was consistent with the Zank21 model. Interestingly, the ratio R_{OT} for the simulation data is lower than unity, which is consistent with our findings.

In conclusion, we showed that the transmission of the magnetic field fluctuations across IP shocks can be successfully modeled by the Zank21 model. We suggest that the model can predict the downstream levels of fluctuations even in scenarios where the assumptions of the theory are not well met. We show that the variability of the solar wind within the inertial range of turbulence effectively prohibits accurate determination of the downstream levels with a good accuracy. We have found that any particular prediction of downstream PSDs suffers an error of a factor ~ 3. We plan to extend our study where we will address the transmission of velocity and density fluctuations across IP shocks and terrestrial bow shock, employing the methodology described above.

Acknowledgments
This work is supported by the Czech Grant Agency under contract (23-06401S). This paper uses data from the Heliospheric Shock Database, generated and maintained at the University of Helsinki. G.P.Z., M.N., L-L.Z, and L.A. acknowledge the partial support of a NASA Parker Solar Probe contract SV4-84017, an NSF EPSCoR RII-Track-1 Cooperative Agreement OIA - 2148653, a NASA IMAP subaward under NASA contract 80GSFC19C0027, and a NASA award 80NSSC20K1783.

References

[1] Zank G P, Li G, Florinski V, Hu Q, Lario D and Smith C W 2006 *Journal of Geophysical Research: Space Physics* **111** (*Preprint* https://agupubs.onlinelibrary.wiley.com/doi/pdf/10.1029/2005JA011524) URL https://agupubs.onlinelibrary.wiley.com/doi/abs/10.1029/2005JA011524
[2] Borovsky J E 2021 *Frontiers in Astronomy and Space Sciences* **8** 5
[3] Pitňa A, Šafránková J, Němeček Z, Ďurovcová T and Kis A 2021 *Frontiers in Physics* **8** 654
[4] Luttrell A H and Richter A K 1987 *Journal of Geophysical Research* **92** 2243–2252
[5] Zank G P, Li G, Florinski V, Hu Q, Lario D and Smith C W 2006 *Journal of Geophysical Research (Space Physics)* **111** A06108
[6] Hu Q, Ao X, Peltzer R and Zank G P 2012 Turbulence properties associated with interplanetary shock waves *Space Weather: the Space Radiation Environment: 11th Annual International AstroPhysics Conference* (*American Institute of Physics Conference Series* vol 1500) ed Hu Q, Li G, Zank G P, Ao X, Verkhoglyadova O and Adams J H pp 192–197
[7] Adhikari L, Zank G P, Hunana P and Hu Q 2016 *Astrophysical Journal* **833** 218
[8] Adhikari L, Zank G P, Hunana P and Hu Q 2016 The interaction of turbulence with parallel and perpendicular shocks *Journal of Physics Conference Series* (*Journal of Physics Conference Series* vol 767) p 012001
[9] Borovsky J E 2020 *Journal of Geophysical Research (Space Physics)* **125** e27518
[10] Pitňa A, Šafránková J, Němeček Z, Goncharov O, Němec F, Přech L, Chen C H K and Zastenker G N 2016 *Astrophysical Journal* **819** 41
[11] Zank G P, Nakanotani M, Zhao L L, Du S, Adhikari L, Che H and le Roux J A 2021 *Astrophysical Journal* **913** 127
[12] McKenzie J F and Westphal K O 1969 *Planetary and Space Science* **17** 1029–1037
[13] Zank G P, Dosch A, Hunana P, Florinski V, Matthaeus W H and Webb G M 2012 *Astrophysical Journal* **745** 35
[14] Zank G P, Nakanotani M, Zhao L L, Adhikari L and Telloni D 2020 *Astrophysical Journal* **900** 115
[15] Adhikari L, Zank G P, Telloni D, Hunana P, Bruno R and Shiota D 2017 *Astrophysical Journal* **851** 117
[16] Kilpua E K J, Lumme E, Andreeova K, Isavnin A and Koskinen H E J 2015 *Journal of Geophysical Research (Space Physics)* **120** 4112–4125
[17] Hu Q and Sonnerup B U Ö 2002 *Journal of Geophysical Research (Space Physics)* **107** 1142
[18] Zank G P, Adhikari L, Hunana P, Shiota D, Bruno R and Telloni D 2017 *Astrophysical Journal* **835** 147
[19] Lepping R P, Acũna M H, Burlaga L F, Farrell W M, Slavin J A, Schatten K H, Mariani F, Ness N F, Neubauer F M, Whang Y C, Byrnes J B, Kennon R S, Panetta P V, Scheifele J and Worley E M 1995 *Space Science Reviews* **71** 207–229
[20] Ogilvie K W, Chornay D J, Fritzenreiter R J, Hunsaker F, Keller J, Lobell J, Miller G, Scudder J D, Sittler Jr E C, Torbert R B, Bodet D, Needell G, Lazarus A J, Steinberg J T, Tappan J H, Mavretic A and Gergin E 1995 *Space Science Reviews* **71** 55–77
[21] Torrence C and Compo G P 1998 *Bulletin of the American Meteorological Society* **79** 61–78
[22] Dudok de Wit T, Alexandrova O, Furno I, Sorriso-Valvo L and Zimbardo G 2013 *Space Science Reviews* **178** 665–693 (*Preprint* 1306.5303)
[23] Pitňa A, Šafránková J, Němeček Z, Franci L, Pi G and Montagud Camps V 2019 *Astrophysical Journal* **879** 82
[24] Efron B and Tibshirani R J 1993 *An Introduction to the Bootstrap* (*Monographs on Statistics and Applied Probability* no 57) (Boca Raton, Florida, USA: Chapman & Hall/CRC)
[25] Zhao L L, Zank G P, He J S, Telloni D, Hu Q, Li G, Nakanotani M, Adhikari L, Kilpua E K J, Horbury T S, O'Brien H, Evans V and Angelini V 2021 *Astronomy & Astrophysics* **656** A3 (*Preprint* 2102.03301)
[26] Trotta D, Pecora F, Settino A, Perrone D, Hietala H, Horbury T, Matthaeus W, Burgess D, Servidio S and Valentini F 2022 *Astrophysical Journal* **933** 167 (*Preprint* 2202.14029)
[27] Rakhmanova L S, Riazantseva M O, Zastenker G N and Yermolaev Y I 2022 *Universe* **8** 611
[28] Nakanotani M, Zank G P and Zhao L L 2022 *Astrophysical Journal* **926** 109 URL https://dx.doi.org/10.3847/1538-4357/ac4781

AUTHOR INDEX